THE SUPERLUMINAL UNIVERSE

T0244095

THE SUPERLUMINAL UNIVERSE

Redefining Consciousness, Time and Space

Régis Dutheil and Brigitte Dutheil

WATKINS
Sharing Wisdom
Since 1893

ACKNOWLEDGEMENTS

Our warmest thanks to Joanne Esner of Éditions Sand, whose dedication to and enthusiasm for the ideas contained in this book have been a great help to us.

Our gratitude is likewise extended to Professor Stuart Edelstein, Director of the Biochemistry Department at Geneva University, who was heavily involved in the elaboration and the shaping of this work.

We would also like to acknowledge several other colleagues at Éditions Sand, in particular Carl van Eiszner, Frédéric Ferney, Agnès de Gorter and Roger Vinciguerra. Each one of them has contributed in a constructive manner to the realization of this project.

R and B Dutheil

This edition first published in the UK and USA in 2024 by
Watkins, an imprint of Watkins Media Limited
Unit 11, Shepperton House
89-93 Shepperton Road
London
N1 3DF

enquiries@watkinspublishing.com

Text copyright © Éditions Sand, Paris, 1990
Design and typography copyright © Watkins Media Limited 2024

Translated into English by Matt Raymond

Régis Dutheil and Brigitte Dutheil have asserted their right under the
Copyright, Designs and Patents Act 1988 to be identified as the authors
of this work.

1 2 3 4 5 6 7 8 9 10

Typeset by JCS Publishing Ltd.

Printed and bound in the United Kingdom

A CIP record for this book is available from the British Library

ISBN: 978-1- 78678-879-5 (Paperback)
ISBN: 978-1-78678-886-3 (eBook)

www.watkinspublishing.com

CONTENTS

INTRODUCTION

From time to time, we all stop and question who we are, what we're doing and the world around us. This often involves the sort of timeless questions that require timeless answers, because they influence the lives we lead. Perhaps it concerns even more than this, as questions about life are inextricably intertwined with those of mortality, our human limits and the inevitability of death, which we've been told is the end. However, such abiding questions often go unspoken, relegated to moments of deep introspection, and sometimes muted by the sheer rawness of the revelations themselves: those silent, everyday moments when our mind wanders, when we stare off into space at work or when fleeting ideas pop up spontaneously in the presence of familiar ones or new acquaintances. In these instances, our innate responses take a certain precedence, as they construct the framework through which we perceive and feel, and thus everything that we are – on various different levels – conscious of.

Indeed, the subject of consciousness rather finds itself in the spotlight, which highlights the importance of pursuing its exact nature. The great thinkers of our most ancient civilizations were aware of this, and we see the power of their convictions reflected in how they inspired the masses to define consciousness in ways that coincided with their own interpretations – which can be quite diverse, although underlying patterns do emerge. We find, for example, that a certain level of coherence regarding a society's understanding of the conscious experience is psychologically necessary in order to limit the universe it occupies. Here we also get the first hint of another connection – between theories that model consciousness and those that model the universe, which is all the more compelling if one adopts the perspective that the universe only exists for an individual to the extent that they are conscious of it.

The spotlight on consciousness is reflected in language itself, with a host of everyday expressions passed down from generation to generation (*conscious of, losing consciousness, regaining consciousness*) that suggest the latent comprehension of connected concepts in close correlation with the conscious state (*existence, reality, truth, known, unknown*) – each of which possesses an abstract nature that in turn invites its own examination in relation to the evolution of human understanding. This is because what we know (or *think* we know) and what we perceive from our surroundings fundamentally determine our choices and actions, as well as the evolutionary path that society chooses to follow. Our interpretation of the universe and reality – perceived through forcibly biased senses – feeds and sculpts our conscious being, and it's the very theories through which we understand this conscious experience that create the foundations for our understanding of the reality that surrounds us.

The goal of the modest treatise in your hands is to inspire advances in the human understanding of the conscious mind. To that end, you are about to find yourself embarking on a tour through radical philosophies spanning from antiquity to the present day, touching on the fascinating science of modern relativistic physics and quantum mechanics, and embarking on extraordinary expeditions through personal accounts and medical analyses of near-death experiences. You'll dive into the latest scientific interpretations of sensory information, and travel back in time to discover a treasure trove of spiritual concepts that transcend religious borders.

Our goal is to inspire the synthesis of scientific and non-scientific perspectives, so that key findings may be extracted from a broader body of research and ideas pursuant of the same objective: evolving our understanding of the conscious mind to something closer to its true nature.

By way of an example, we will throw our own theory into the mix along the way, in the hopes of nurturing and expanding your own perception of matter, reality and in particular spacetime. We will apply our schema to Einstein's theory of relativity, developing a new interpretation of past/present/future simultaneity, all the while proposing a new physical paradigm of the conscious experience that brings us back to the heart of our ancestors' teachings.

Synthesizing these approaches in this way might very well prove vital to the advancement of a human understanding of the conscious mind, and we see this reflected in the desires and intuitions concerning a perpetually elusive reality revealed in the works passed down to us from Pythagoras, Plato, Democritus, Leibnitz, Descartes and Hegel; the modern scientific discoveries of Einstein, Kammerer, Jung and Pribram; in the analyses of near-death experiences carried out by Kübler-Ross, Moody, Ring and Sabom; and in the research into reincarnation by Stevenson.

The broad-minded willingness to consider new concepts – such as new theories of matter, a version of time that doesn't work like we thought it did, a death that *displaces* instead of *disappears*, and the presence of humanity's history within each and every one of us — represents the act of reaching, and the possibility of grasping, another dimension of reality. Such insights expose the true nature of the human condition: a timeless greatness, perpetually reaffirmed by the smallest of acts.

PART ONE

1
CONCEPTUALIZING CONSCIOUSNESS

Every morning we wake up. Sometimes, we may find ourselves returning from a distant daydream. Occasionally, we may even come to after passing out or from anaesthesia. Our daily habits make it all seem rather simple: we notice the light, perhaps take a note of the ambient temperature, and then our sensory processors kick into gear and we recap our memories.

The entire spectrum of these experiences, from instantaneously waking up to the more prolonged emergence from an anaesthetic, gives us the impression of "returning to ourselves", like reintegration into a vessel which has been temporarily abandoned for the obscure shores of the unconscious. What was blurred becomes focused, and we sense that there's a psychological need to locate ourselves both spatially and temporally. The question invariably posed by patients who wake up in a hospital bed is: "Where am I?"

The Nature of Consciousness

What is consciousness? The pervasiveness of this question throughout centuries of philosophical inquiry gives some insight into the nature of the mystery at hand. A modern definition might sound like:

The perception, to a certain degree of clarity, of phenomena that provide sensory information pertaining to one's own existence.

However, this is not immune to a dependence on certain abstract concepts that elicit further enquiry in turn, such as *existence* and *reality*.

Another approach might be to define consciousness as the union of sensory information and the intelligence that analyses it and interprets meaning. However, this manoeuvre carries the additional responsibility of establishing a primordial distinction between an entity and the physical world, while inherently rendering the conscious and its associated physical form inseparable.

Whatever the case may be, the schema through which we interpret the nature of consciousness plays an important role, as it serves to direct the investigations we choose to pursue.

Philosophical Perspectives

Historically, philosophical approaches to consciousness have been very varied. However, we can begin to divide and conquer these by establishing a spectrum that ranges between two extremes:

- **Homogenous:** homogenous approaches are characterized by the predominance of a single theory. For example, a completely materialistic outlook that considers all forms of spirituality an illusion, or conversely, total spirituality that endows all bodies with a spirit.
- **Heterogeneous:** this end of the spectrum involves schemas that are more multifaceted (in practice, there's usually a duality); for example, a theory of the universe that allows spiritual aspects to differ completely from those of the material realm.

Homogenous Perspectives

To begin our investigation into the nature of consciousness, we're about to take a short stroll through history, from Ancient Greece to the 20th century.

Democritus

Unitary materialism is a philosophy that predates Socrates and leads us all the way back to the famous Ancient Greek philosopher and scientist Democritus (520–460 BCE).

His was a time when philosophy and scientific research were deeply intertwined, to the extent that most famous thinkers of the era regularly employed combinations of these methods (perhaps a bit eccentrically in some cases), in particular with respect to solving one particularly challenging question to which they gave great precedence: the origin of the universe.

Democritus, whose original writings have proven all but lost, is widely considered to be the forefather of *physical materialism*. The scientific foundation of the schema he proposed seems incontestable and is relatively straightforward: each material body is composed of atoms, which are the smallest units in which matter can exist (and invisible to the human eye), and are indivisible, indestructible, eternal and solid. Democritus postulated that atoms fall freely in a vacuum and travel at very high velocities, but would follow trajectories that weren't necessarily straight — whereby he attributes the formation of compounds, and therefore all matter and spiritual bodies, to the inevitable collisions of these elementary particles. With the benefit of 2,000 years of hindsight, we can see that Democritus actually laid the framework for early 19th-century atomic theory.

But back in antiquity, his proposal was not only revolutionary and unique, it reconciled two popular opposing philosophical doctrines during the 6th and 5th centuries BCE:

- **Heraclitus of Ephesus:** Heraclitus and his followers adhered to a schema of the universe that emphasized a constant state of flux, forever changing: "No one ever steps into the same river twice."
- **Parmenides and Zeno of Elea:** this school of thought, on the contrary, believed that mutability was an illusion, a trick of the senses, and that what was *real* was instead characterized by what was immutable, unique and eternal.

Democritus's solution was very bipartisan, consigning atoms to the immutability camp and the compounds they form to the mutable party. The result: a universe in a constant state of flux, evolving and decomposing, whose fundamental construction was based on elements that were both indivisible and indestructible. This also inadvertently implies that nothing is born from or returns to the void, which was considered a very risqué proposition at the time on account of its proximity to challenging the existence of a divine power over creation and death.

The theory also defines the soul (which we can interpret as the conscious mind) as the construct of a special class of atoms capable of penetrating physical bodies and reaching their interiors. In this way, the soul and body are considered permanently entwined, the soul permeating the whole of the body and forming a nucleus in the heart where it supplies "the breath of life". The soul is considered mortal, based on its observed disappearance and dissolution along with the body at the time of death, at which point its constituent atoms re-join the eternal flux.

Democritus's theory can be considered *homogenous* in the sense that it more or less perceives the entire universe through a single atomic schema, and *materialist* as well, as it gives both the material and spiritual (therefore its entire scope) this same construction.

Atomic Schools

Democritus founded a school that flourished for several centuries, whereupon Epicurus (340–270 BCE) took up the torch, refined certain philosophical arguments and created a more complete version of Democritus's theory, which became known as Epicureanism. None of his original work survives today, save for a few maxims and three letters. But at the time the breadth of influence of Epicurus's Athenian school was great enough to reach Rome, where it was adopted by the Latin poet Lucretius (98–55 BCE) with the express intent of making it readily available to the public. The result is a lengthy six-book philosophical poem entitled *On the Nature of Things*.

So passed the schema's indoctrination into literature, and the vast phenomena of the universe (including perception and ideas) into the workings of a great mechanism. Lucretius postulated that sensory perception was induced by miniscule, invisible corpuscles called *simulacres*, which inhabited the atmosphere. Upon contact with sensory organs, they would induce various effects on the spirit. Lucretius also introduced a taxonomy of these, whereby certain simulacra emanated from objects' surfaces or interiors, according to the role they would play in vision, sound, taste and smell. Others, like those constituting ideas, would form spontaneously in the air, sharing an atmospheric pervasiveness with their sensory counterparts, but with a corpus so minute that they were able to pass through the very pores of the body.

Against the backdrop of antiquity, Epicureanism stands out for its uniqueness and originality: during an epoch notorious for a lack of separation between religion and state, it wasn't standard practice for a philosophical school to promote ideas that denied the existence of the gods and promoted science as the only source of hope and progress for humanity. However, the religion of the time probably fell somewhat short in terms of what it could deliver. Belief systems that attributed thunder and eclipses to notions of divine punishment inevitably fostered anxiety and ignorance, while paralysing courage and reason. At the same time, debunking divine interference to the status of simulacra probably served to loosen the grip of religion, as well as cultivate objectivity and rationality regarding natural phenomena.

The miraculous survival of Epicureanism can also be attributed to the democratic system in place in Athens, as well as the relative liberty (even audacity) of Greek citizens with regard to religion. However, the subsequent and rather rapid fall of Greek culture from the limelight, combined with the Church's sway over science and culture for the next thousand years, are likely culprits in materialism's stagnation until the Renaissance. We won't see it flourish again until the 16th century, and the 18th and 19th centuries in particular.

During the hiatus, the functional need to separate philosophy and theology, and to promote an experimental methodology, gradually imposed itself upon philosophers and scientists. Francis Bacon (1561–1626) became one of the first to emphasize the importance of an experimental method and the rational observation of natural phenomena. He also offered a new, rather optimistic interpretation of science's ultimate role: to extend the power of the human race indefinitely.

Bacon's ideas would have profound influences on René Descartes (1596–1650), and to a larger extent the English philosophers Thomas Hobbes (1588–1679) and John Locke (1632–1704). Hobbes, first and foremost a man of politics, was partial to a totalitarian state where force took precedence over law, drawing on elements of Bacon's *empiricism* as well as the *atomism* of Epicurus and Democritus to derive a rather pessimistic interpretation of society: the proverbial *Homo homini lupus est*, or "humans are wolves to their own kind".

For Hobbes, the senses lay at the root of all knowledge and constitute the very first principles we become acquainted with. For him, the imagination was a sort of unedited grouping of sensory parcels, memory a reflection of former sensory experiences, and ideas a by-product of the materialistic spirit. His schema thus involves a complete mechanization of the mind, with the heart acting as a sort of watch spring and our sense of morality originating from the instinct for preservation, affirmation and power.

The schemas proposed by Locke and the French philosopher Condillac (1715–1780) place a similar focus on the senses. Locke believed that all complex ideas are comprised of simpler forms that one acquires through experience. He wasn't strictly materialist, because he recognized the existence of God, but his ideas are widely cited in the doctrines of materialism. Condillac published a *Treaty on Sensations*, which similarly reduces spiritual functions to sensory-based interactions. He illustrates a particular thought experiment involving a statue, postulating that the ability to endow it with a particular sense, smell for example, would be the first step in reconstructing the human psyche. If the statue inhaled the scent of roses, for example, that scent would constitute its entire experience, which Condillac suggested would furnish the key to its attention, while the remnants of the odoriferous particles would give way to memory, and desire would awaken from the need to relive the enjoyable experience.

Marx: Materialism Reaches Historic Heights

In the 17th and 18th centuries, the doctrines of empiricism and sensationalism furnished the foundations for the materialism of the German philosopher Ludwig Feuerbach (1804–1872), who in turn inspired Karl Marx (1818–1883). It's the phosphorous within, Feuerbach argued, that does our thinking for us: a bold claim that nevertheless follows its own careful rationale, perceiving the material properties of the brain as the ultimate source of every component of the human conscious experience (including the idea of God), whereby the spirit is then simply a reflection of the material conditions that produce it.

Karl Marx used Feuerbach's ideas as a launch pad for *historic materialism*, a sort of amalgamation of Hegelian principles (such as the primacy of history, progression through the resolution of contradictions, etc.). Marx stated: "It is not the consciousness of men that determines their being, but, on the contrary, their social being that determines their consciousness."

The materialistic schema he proposed places a special emphasis on one law in particular: equal and opposite reaction. He believed that if people are considered as a product of the environment in which they live, then conversely we should assume them capable of acting reciprocally on that environment and transforming it through work. This last point then brings us to the *historic* in Marx's *historic materialism*, as the doctrine identifies the source of societal evolution as the forces behind the advancement and optimization of production techniques. In other words, his schema emphasized the influence of industrial technology (for example, the windmill during the Middle Ages, or the 19th-century steam engine) on societal structure (feudalism and capitalism in these instances). Society's infrastructure is then perceived through the lens of productive forces and relationships (the latter representing relationships mandated by the current economic/technological state), while ideas around justice, philosophy, religion and art are relegated to by-products.

20th-Century Materialism: Enter Psychology and Biology

The 20th century saw biologists and psychologists hard at work constructing elaborate architectural paradigms of the conscious mind, and materialism flourished. It saw the birth of behaviourism, for example, which conceptualizes the conscious experience into more or less mechanized patterns of behaviour conditioned in response to physical stimuli, whose relationships can be studied experimentally. Subjectivity then effectively takes a back seat, allocated to the status of a secondary effect controlled by the brain and nervous system.

More recent paradigms come from the field of cybernetics, where the temptation has proven irresistible to play off functional analogies between computers and the human brain. Or tortoise brains, rather, as 1950s England witnessed the birth of cybernetic tortoises: the progeny of the neurophysiologist and robotics pioneer William Grey Walter (1910–1977). They were considered autonomous robots equipped with conditioned reflexes analogous to those of living creatures. For example, the tortoises could recharge themselves when their energy levels became critically low. The extrapolation of these ideas then naturally emboldened the analogy, leading some experts to suggest that the underlying mechanics of life forms would inevitably be revealed through robotics and physiochemistry.

One can then wonder: what role does the conscious mind play in this brave new world?

- In *Design for a Brain* (1952), W R Ashby simply states that he won't make any mention of consciousness or the subjective concepts it entails, because introducing these concepts isn't necessary for his work.

- In *The Modelling of Mind* (1963), MacKay describes consciousness as a collection of organic elements reacting across an information network.

Take a Walk on the Spiritual Side

On the flipside of the homogenous coin, we have schemas that employ a purely spiritual approach, which are of significantly less variety (in the broadest sense) than their materialist and dualist counterparts. However, the second half of the 17th century saw notable efforts on the part of two famous philosophers.

Baruch Spinoza (1632–1677) developed his theories from the Cartesian doctrine and constructed a pure form of *pantheism*: the belief that the universe is entirely embedded in God, who takes the form of a pure and self-sufficient substance. Spinoza's definition in his *Ethics* states that God is "a being absolutely infinite, that is, a substance consisting of an infinity of attributes, of which each one expresses an eternal and infinite essence". However, of these infinite attributes, only two would be within the grasp of human understanding: *thought* and *extension*, whose properties, or rather collections of them called *modes*,[1] defined both the world and human beings, while free will was reduced to an ignorance of the factors that determined it.

We see many similarities with Spinoza's theories in the work of Gottfried Leibniz (1646–1716), but although the two even met once, they enjoyed a remarkably different quality of life. Spinoza was expelled from Amsterdam's Jewish community when he was 23 years old after his ideas were judged heretical. He earned a relatively modest salary fabricating telescope lenses and suffered an early death at the age of 45. Leibniz, on the other hand, lived to be 70 and was an important trusted political advisor in several German states. He founded the Berlin Academy, independently discovered calculus at the same time as Newton, and held a high rank in a secret occult society known as the Rosicrucians. Leibniz showed an aptitude for integration and interpersonal skills and was particularly keen on developing a more complete version of pantheism than Spinoza had envisioned. On the side, he had a secret project to discover a "universal combinatorics": a sort of philosophical calculus that would allow one to determine any truth from a set of rigorously defined elementary symbols.

The levels of idealization and homogeneity in the works of Spinoza, Leibniz and Descartes, as well as in the empiricism of Hobbes and Locke,[2] seem a natural product of the success enjoyed by science and the experimental method after the Renaissance had cleared away certain highly philosophical remnants of the Middle Ages. There is also the impression that the 17th century's propensity for absolute monarchies somehow spilled over into its philosophy. Only time will tell whether this epoch's claim on the two most original and idealist philosophical doctrines will remain intact.

Leibniz certainly hit a high note in that regard with his 1714 publication of *The Monadology*, in which he suggested that the universe is divided into units called *monads*, which represent simple, intangible, active and spiritual substances. Each monad was associated with a form of perception whose quality ranged from humbler varieties (which might lack a memory component), to those more endowed with reason, all the way up to God:

> Monads are the elementary substances from which compounds are formed: that's to say, they have no constituent parts. (...) When there are no constituent parts, the substance is intangible, without form, indivisible and the true atoms of nature. They are elemental.

13

There's an evident lineage connecting this theory to atomism, yet Leibniz's atomic schema is unique: he postulated that atoms are non-spatial, with each one reflecting a microcosm of the universe. Interestingly, he also associated the collective consciousness, or a universal spirit, which operates outside of normal spacetime, with a special master control monad.

The idealized visions of Spinoza and Leibniz enjoyed a much briefer moment in the limelight than their materialist counterparts. However, they are without doubt interesting, and perhaps even more fascinating in part on account of their rarity, than the heterogeneous schemas which pop up much more frequently.

The Dualist Current

Plato's Dualist Universe

As much as homogeneous schemas show a penchant for determinism, their heterogeneous counterparts seem to have a taste for an element of free will. Our first example is Plato (c.429–347 BCE), whose work is rooted in Socrates and other predecessors. The heterogeneity in Plato's schema is drawn rather definitively in the form of a border that separates the *body* and the *soul*: the former being divisible and composed of matter, the latter intangible, indivisible and eternal.

Birth was viewed as a union of the two that would be dissolved at death. For the soul, the cohabitation would signify a loss of purity resulting in its fracturing into three parts: one-part consisting of the higher functions (such as reason and contemplation), destined to maintain order over the two-parts consisting of the lower functions (such as courage, carnal desire and sensory pleasure).

Plato gave the spirit a refuge in the *Realm of Ideals*,[3] a highly original construct based on the general observation that while the world constantly changes and adapts amidst an unquestionable mortality, this also denotes a certain level of consistency, with traits passed down from one generation to the next through reproduction. Plato postulated that this consistency possessed an essence that was both universal and eternal, and he called it an *ideal*. For example, the ideal of a person would be a sort of idealization of personhood, from an evolutionary point of view. This was based on the belief that the designation "being" merits nothing less than an eternal immutability, which eliminated everything except these ideals from the category of that which was alive. That which changed and perished was therefore considered outside the realm of science and investigation.

Each object in nature, even human creations, was thought to originate from an ideal, which was in turn accorded a place in a hierarchical structure that enthroned the ideal of "Good" at the upper limits of the intelligible world. And every single person was believed to carry a piece of the Realm of Ideals within them, a remnant of the spirit's dwelling prior to its association with the physical body and which Plato considered to be the source of reason, while the lower functions were enslaved by the senses. Humans were victimized, in a way, falling prey to sensory illusions that merely refracted truth.[4] Upon death, the spirit would return to the Realm of Ideals and reincarnate regularly throughout the ages.

Plato's schema shows a lot of heterogeneity at work, in particular duality: a world of senses and images that governs the body and a Realm of Ideals for the soul, with both components having a completely independent existence.

Aristotle: The Soul and the Body – Two Faces of the Same Reality"

As Plato's student, Aristotle was naturally inclined to raise some objections, in particular to the rationale concluding that the soul possesses the ability to take up residence in just any old body. However, regulating this phenomenon would require passing the required legislation, which Aristotle achieved through the re-designation of the soul as an *entelechy*: the debut of an organic body that has the potential for life.

In *Physics and Metaphysics*, he postulates two fundamental qualities with complementary roles that he calls *matter* and *form*:

- **Matter: undetermined potential.** Aristotle perceived matter as compounds that lacked natural units – something that doesn't exist in and of itself and that's incapable of acting without being animated by an exterior agent. Material bodies would then be rather like dormant machines, ready and waiting to go, but lacking what's necessary to start up. In a word, *potential*.
- **Form: the realizer of potential.** Form is what realizes matter's potential, i.e. it is the soul that animates the body (the first entelechy of the body). Form itself is notably homogenous; that is, it is not a compound assembled from constituent parts.

These complementary roles are used to remove any claim that the soul can freely inhabit other bodies, by identifying each person as the product of a binding union between a soul (form) and a body (matter).

In some ways, the use of categorical abstractions renders this schema even more idealized than Plato's. We can also see a natural heterogeneity develop (again, duality), where each component shares the effects experienced by the other, in particular at death. However, Aristotle was the son of a medical doctor and had more of a taste for the observational, which frequently put him at odds with Plato, who tended to favour the mathematical and theoretical. This was also an era in which the various strands of science didn't each strain against each other, pulling in their own directions, and Aristotle was similarly a well-rounded individual, dabbling in everything from politics to theatre, physics, meteorology, biology, logic and metaphysics. He seemed hell-bent on analysing the nature of the universe and its inhabitants.

Aristotle's theory can also be seen in a utilitarian light, as providing the means to an observational end, since we arrive at a schema in which the only reality of concern is what's concrete and observable. Plato's ideals, on the contrary, inherently relegated objectivity away from observable phenomena. For example, while Plato might primarily associate an offspring with the quintessential ideal of its species, Aristotle would perhaps prefer to attribute its traits solely to its parent.

However, the idea of a perpetually evolving world is present in both schemas, as it is in those of Parmenides and Zeno of Elea,[5] and Democritus's atomism. The idea of capturing the irrefutable coexistence between mutability and immutability proved a prized target in the eyes of ancient Greek philosophers, with Democritus attempting to do so through his immutable elementary particles, whose compounds were in constant flux, and Plato through an immutable Realm of Ideas that served as a basis for modelling our everyday changing reality.

Aristotle stalked his prey more stealthily, though, subtly ensnaring it by drawing a line between *action* and *potential*, using matter's own penchant for variation against it: for example, an acorn is preordained to become an oak tree, but the *exact* outcome of this potential depends on the *action* that grows it, and the *being* that sees it through.

Aristotle postulated that the soul acted as a sort of a broker between potential and action, resourceful enough to navigate the four causes that he believed to characterize all phenomena in the universe:

1 **Material**: what something is made of (e.g., marble is the *material cause* of a statue).

2 **Form**: the driving force behind development (for the statue, perhaps the idea imagined by the sculptor; for a human, the soul).[6]

3 **Agent**: the direct antecedent that provokes the change (e.g., the chisel strike that forms the sculpture).

4 **Endgame**: the ultimate goal envisioned (for the sculptor, this could be fame and fortune).

Aristotle suggests that nature itself determines the ultimate endgame, thereby guiding each life form toward a purity that represents its most perfect realization, which he associates with God, whom he believes to be the first and final cause, within whom everything will be realized and perfected.

From Rationalism to the Birth of Science

Aristotle is also often considered a founder of *rationalism* on account of his work with chains of logic and categorization, which paved the way toward scientific developments. Furthermore, Aristotle's schemas proved flexible in practice (much more so than their idealistic Platonic homologues, which fought feverously to preserve their untarnished perfection), and could be employed toward diverse ends, after which point the results could be reframed in the original schema.

The fall of the Roman Empire took Aristotle's texts with it, and it wasn't until the Middle Ages (notably after the 12th century) that Western Catholicism discovered the only copies of his work had been preserved in Arabic cultures. The resulting philosophic reawakening rapidly gained enough momentum to sound alarm bells in the Catholic Church, who responded by assimilating it into their own teachings. The 13th century, for example, saw St Thomas Aquinas publish a synthesis of Aristotelian naturalism and Catholic law entitled *Summa Theologica*.

From there, the remnants of Aristotle's teachings would spend the next centuries on a rather downhill slope into an academic dogma that would incite significant insurrections during the Renaissance (through the work of Bacon and Copernicus) and the 17th century (culminating with Galileo, Kepler, Newton and Descartes).

Descartes: Founder of Scientific Rationalism

Like Aristotle, René Descartes had a multidisciplinary background, yet he wasn't satisfied with the education he acquired at several universities, so he decided to teach himself. In 1619, the then 25-year-old Descartes dreamed that he was destined to invent a science that would unite all human knowledge, and all his publications between 1628 to 1648 seem dedicated to this pursuit.

He built a methodology based on four principles:

- evidence
- analysis
- synthesis
- enumeration

From this, he developed a schema of the universe by way of deductions concerning a thought experiment premised on God's existence:

> With everything there is a doubt, save the existence of God, because God is perfect, and perfection implies the reality of one's existence. God is the creator: all truths were created by God's will; God creates each moment in the world that surrounds us.

Nature was thereby stripped of the power accorded it by the Aristotelian schema, and its inherent control of the endgame was reduced to a mechanical transparency that

Descartes predicted would be completely explicable through mathematical reasoning. Other Aristotelian concepts (such as *form*, *action* and *potential*) naturally followed suit, and were struck down by the same blow; so too was the idea of nature worship (by excluding the idea of nature as a goddess). Ultimately, Descartes laid the metaphysical foundations for scientific rationalism, whereby science would gain a sort of independence, and humans a free will accorded at each instant by God. The mechanization instilled in this perception of nature notably led Descartes to consider animals as soulless, resulting in the notorious "animal-machine theory" which had a notable influence on Hobbes's "mechanism".

Descartes's schema bears some resemblance to Plato's but is perhaps further along the heterogeneous spectrum than Aristotle's: body and soul were no longer subcategories of logic and forms, but, rather, distinct realities that maintained a close proximity throughout. As a substance, the soul was thought to be pure, indivisible and immortal, in contrast to the composite and divisible body which it had the capacity to animate. The soul constituted the base of self and thought, action, the will of the passions, and a means for escaping the corrupt nature of the material world, and was generally synonymous with the act of being self-conscious in the sense of distinguishing oneself from the physical world. (Pre-empting the terminology of our next proposition, Descartes's schema can be perceived as introducing two distinct forms of matter, each obeying its own spatiotemporal laws.)

Descartes also gave the soul a physical location within the body: not the heart or brain per se, but a small gland within the brain, which he postulated was an intermediary that gave the soul control over the body.[7] This idea would later be revived by 20th-century neurophysiologists such as Australia's John Carew Eccles (1903–1997): according to Eccles, the spirit is a material reality that, through yet to be detected physical influences, communicates with the brain via single neurons, which produce chain reactions throughout the cerebral cortex.

Kant: A Premise for Relativistic Physics

Immanuel Kant (1724–1804) also took a turn at associating thought with the soul and consciousness, and additionally drew clear distinctions between what he termed *sensibility* and *reason.*

Sensibility was taken to be the means through which individuals accessed the world of sensory phenomena. Spacetime was subject to intuitive interpretation, which then required one to exercise sensibility. Fundamentally, this limited sensory capacity to the interpretation of representations, which constitutes a stricter framework for the interpretation of phenomena. Reason, on the contrary, which more closely resembled the idealism of the spatiotemporal domain, was gifted with the special ability to access an innate understanding of objects that Kant described as *noumena* (things in themselves), which were at the same time free of sensory attributes and thus inaccessible to the senses, although they formed a basis for the interpretation of sensory information.

The idealism of Kant's segregation of the self from phenomena approached Platonic levels. He established a hierarchy of reason with God at the helm, whose comprehension he suggested was impossible outside the realm of pure reason, and he employed Cartesian logic not only in associating thought with consciousness, but in a particular brand of spatiotemporal categories that would afford a sneak peek into the strange quantum and relativistic findings that would occur more than a century later.

Hegel and the Development of the Conscious Mind

Georg Wilhelm Friedrich Hegel (1770–1831) abandoned Kant's radical distinction between mind and body and proposed a three-staged schema of the development of the conscious mind: *thesis*, *antithesis* and *synthesis*.

This began with the *absolute*, which was pure thought and immaterial. Pure thought's dissolution through spacetime then embodied nature, and upon return from this pilgrimage would acquire a certain self-awareness and independent purpose, thereby becoming the conscious mind. Hegel extrapolated this idea to a perception of history as the development of a universal spirit over the course of time.

A New Paradigm for the Conscious Mind

Through these tempestuous seas of philosophical inquiry we can begin to see that, regardless of what type of ship one chooses to set sail in, it's necessary to navigate the wayward channel of the conscious experience prior to plotting a course to distant shores. Indeed, these issues are inherently integrated into the hull of the ship, and this overall structure will be the subject of our next chapter.

We'll be launching our own schema alongside this, in an effort to reconcile the great schemas of our predecessors, along with various other experimental observations, analyses and forms of documented reports. Our newly acquired vocabulary can help to outline the approach: we'll employ a heterogeneous, materialistic schema that will model the conscious mind as the product of a special form of matter known as *superluminal matter* (i.e. matter travelling faster than the speed of light). Modern physics suggests that superluminal matter experiences a spacetime that's drastically different from our own *subluminal* (less than the speed of light) version, and possesses a uniquely distinct *entropic arrow* (more commonly known as "the arrow of time"), which we'll consider through the lens of "information". The duality of our schema will naturally be reminiscent of many of the philosophies we've encountered so far, and in particular the Platonic distinction between the Realm of Ideas and that of shadows.

2
CONSCIOUSNESS AND REALITY

The definitions of the conscious experience and reality are interdependent and inseparable: the former is effectively the means by which we know the latter. As a result, this likely couple finds itself seated at the head of most of history's great philosophical schemas, regardless of where they land on the spectrum: a paradigm of the conscious mind inadvertently breaks ground on a schema through which it understands reality, and vice versa.

This can be rather perplexing, as in our everyday lives reality seems inarguably self-evident. However, if asked to define the terms "real" and "reality" beyond dispute, it becomes clear that the self-evident can possess a talent for abstraction that eludes definition.

It's perhaps somewhat problematic to slide through life based on such vague everyday conceptualizations that delimit the surrounding universe, because, in the end, the world is not so very different for us as it was for our ancient ancestors: that is, a vast enigma. Rather, the only difference is that in the present we imagine we've figured it all out (or are just about to), and don't give a second thought to reaffirming this attitude by name-dropping some choice technical terms which, when it comes down to it, we actually understand very little about.

What is Reality?

Finding ourselves between a rock and a hard place, we can fall back on practicalities and opt to define reality based on what we can see and touch: in brief, the sensory information we interpret from our surroundings. Physically, organisms receive sensory information from their immediate environment, and their interpretation and analysis of this information then characterizes their conscious experience. A dictionary might be cautious about associating the notion of "truth" with this, but this passes the buck from one abstraction to another, while etymology would bring us to the Latin *res* ("thing"), which is sort of a redirect toward the observational/sensory approach.

In any case, we will find a repeat of our experience so far with schemas, and the trail of definitions quickly circles back to the conscious experience. The issue is that this presupposes a certain level of consistency and continuity in our observations, let alone the interpretation of sensory information, worthy of evincing our inherent association of *reality* with *truth*.

Reality's Gone and Isn't Coming Back

Fundamentally, we can break down the conscious experience into:

- sensory information
- the processing of this information through the organs (notably the brain)
- the interpretation of this outcome

Whether or not our senses are actually tricksters (shout out to Descartes) and there's a fault in the brain's interpretation and processing of sensory information that's lying dormant –

biding its time until it wakes and shatters into unfathomable abysses destined to upheave our whole conception of reality – is another matter. Practically speaking, such abysses are more likely to occur along fault lines in processes involved with the collection and interpretation of sensory information. In fact, reality is already perhaps not quite as real as we've been led to believe, but evidently something unmistakably surreal and mysterious.

Take the colour red, for example. Whichever beautiful vibrant colour pops into your head – the defining characteristic that you experience when your eyes fall upon a bunch of roses, for example – really possesses very little in the way of physically tangible reality.

There is a stimulus that elicits a retinal response, thereby transforming into a signal transmitted through the optic nerve and network of neurons en route to the cerebral cortex. Everything's just peachy up to this point, but once inside the cortex the mystery deepens. We can observe a special form of electrical activity via electroencephalograms (EEGs) and other means, and we can assume that the signal arriving in the cortex is carrying all the information necessary to depict the scene at hand. We also can measure the individual's response to the stimuli, which occurs just after said electrical activity.

But there's this tiny disparity, this fleeting moment that occurs between the flurry of electrical activity in the cortex and the interpretation of this phenomenon by the conscious mind into a coherent scene. During that briefest of moments, sensation is born; yet there's nothing in our understanding of electricity and magnetism to predict that it would spontaneously elicit this new type of phenomenon. On the contrary, physically everything behaves as if the chain of events culminates upon arrival in the cerebral cortex, prior to having taken on this new significance. In other words, there is a moment when our perception of red roses exists as an electrical signal carrying encoded visual information that passes through the visual cortex – and that is the homologous limit of what experimental science can tell us about it.

The impasse also seems to suggest that the actual *experience of sensations* is something that resides comfortably outside the realm of observational science: the *sensation of red* you experience is something entirely distinct, having nothing to do with its physical characterization as a band of electromagnetic wavelengths, nor its ability to induce absorption by chemicals or photovoltaic cells. It takes a human being to *see* red where there's nothing more than a wavelength and, in fact, the quantitative properties of electromagnetic waves are likely the furthest thing from that person's mind in that moment – they simply perceive the colour. Consequently, within the brief instant that separates the observed electrical phenomena in the cerebral cortex and the conscious perception of the individual, we suspect something else occurs, something mysterious and quite frankly inexplicable: such is the reality of our sensations.

Reality: the Mysterious Domain of the Senses

Being that the conceptualization of truth depends on sensory interpretation, which as we've just seen seems destined to elude scientific explanation, defining reality itself necessitates an element of caution. There are provisional definitions aplenty, but these quickly get muddied when the situation at hand calls for a more precise look at the transformation and reconstruction of sensory information within the brain, whose output bears little or no resemblance to its input of physical quantities.

The domain of the senses therefore remains, despite many valiant 19th-century attempts to quantify it, categorically subjective: it's evanescent, open to interpretation and to multiple levels of distortion, as it is readily exaggerated by fevers, hallucinogens and a host of other biological and psychological conditions.

Keeping Calm and Carrying On

Being unable to contend with the subjective nature of our perception of reality (and this is without taking into consideration issues introduced by quantum mechanics and relativity, as we'll see later), we're forced back to the start: the conscious experience as a construct of the brain or conscious mind.

We see this reflected in Descartes's decision to centre his schema on the self. Our ship must pass through the channel of conscious experience, and just as we've seen duality arise on more than one occasion (for example, *materialist* versus *spiritual*, *homogenous* versus *heterogeneous*), so attempts over the past millennia to understand consciousness have forged two paths.

The first of these attempts to eliminate as much subjectivity as possible from phenomena with the rather Sisyphean goal of pure objectivity. In fact, this constitutes the foundation of the scientific and experimental methods, where the goal is to sift out universally repeatable measurements and replace the subjectivity of each sense with the objectivity afforded by scientific instrumentation.

The second route is to do an about-face and to endow reality itself with a fundamental subjectivity. Sensory perception and the conscious mind are then reassigned the role of closing the gaps, which inherently extend well beyond their scope and penetrate an even deeper definition of reality.

Scientific Analysis

The hallmarks of the scientific method date all the way back to Aristotle, whose subtle manipulation of chains of logic and categorization has earned him the reputation as one of the forefathers of rationalism. His intent focus on the observational (objects and matter) also resonates comfortably with scientists. However, the groundwork for experimental methodology would have to wait for Galileo in the 17th century.

The Renaissance was followed by a shake-up in the scientific community, as the writing was on the wall for classical Greek and Roman teachings, which had been filtered through the Church during the Middle Ages (for example, that the Earth is the centre of the universe), and taught for year after year. Medicine and astronomy would be among the first to break the mould, validated by their progress and discoveries, which would later serve as foundations for the experimental method. Copernicus and Kepler pioneered the way in basing their research on these principles.

Copernicus broke away from traditional astronomical paradigms at the turn of the 16th century and rekindled the very ancient hypothesis (supported by Pythagoras) of a heliocentric universe. His entirely new paradigm supplanted its Ptolemaic and Aristotelian predecessors, was confirmed experimentally, and was notably credited for its prediction of anomalies observed in the orbit of Venus.

Galileo was a powerhouse of the experimental method. When still only a young man he invented the thermometer and designed a new form of hydrostatic balance. He used his own experimental observations, conducted by himself with a very advanced telescope of his own construction, to discover Jupiter's moons and Saturn's rings, and to derive his own theory of falling bodies: he quickly backed the Copernican paradigm, which correlated well with his own experimental findings.

However, the notorious conflict between Galileo and the Church during the 1610s gave the astronomer the choice of being burned at the stake or spending the rest of his days under house arrest. The plaintiff took issue not only with the Earth orbiting the Sun, but with the experimental method itself, whose emphasis on objective observation directly contradicted certain core ideas in Catholicism (such as faith, divine power and heaven).

Descartes rather circumvented this with a technicality that explicitly restricted the application of his methods to the study of earthly phenomena. This fostered the development of deductive logic, but despite his efforts, the Church still found it necessary to blacklist Descartes' work in 1662.

The use of experimentation and logic nevertheless proved fruitful for the scientific method. However, its ascent to the throne was ironically accompanied by a totalitarian aspect, characteristically seeking the exclusion of anything different. It's important to remember that the scientific method is founded on a rather binary Aristotelian logic (e.g. true or false, something exists or it does not) and is not particularly well suited to handle nuance. As this attitude would only snowball over the centuries to come, the growing absoluteness of its reign revealed a predilection for segregating dissenting voices under the title of "superstition", where they joined the amalgamation of religious perspectives against which science pitted itself mercilessly. Such are the origins of 19th-century rationalism and positivism.

The Experimental Method

The experimental method is essentially about performing experiments under the strictest conditions that one can afford in order to maximize the repeatability and precision of the results, which constitute the principal criteria by which validity is conferred and passed on to theoretical interpretations based on these conclusions. Parapsychological phenomena and the like naturally fall quickly by the wayside.

The method did wonders for physics during the 17th, 18th and 19th centuries. Newton, for example, laid out complete treatises on geometric optics, the wave nature of light, gravitation and mechanics. The ego of science reached a notable high point in the 19th century. Physics reigned supreme over all, and was itself divided into three principal fields:

- Thermodynamics described the transfer of energy and heat.
- Mechanics believed itself to be on the verge of a perfect description of all motion.
- Electricity and magnetism, which also included optics.[1]

The scientists of the time also believed that they'd identified elementary particles (the quintessential atoms, the most tiny and indivisible state of matter envisioned in Democritus' theory and in other materialistic schemas); and the epoch marked a perigee between materialistic schemas of the conscious experience and the objectivity of scientific observation. One 19th-century physicist is even renowned for having stated that he believed that there was nothing left to discover, and that he felt sorry for future generations. Of course, *tel est pris qui croyait prendre ...* which is French for the joke's on you. Cracks and gaps started popping up in different areas, classically symptomatic of faulty foundations; for example:

- The speed of light did not change when observed from different reference frames, which was a direct contradiction of the predictions of classical mechanics.
- Anomalous blackbody radiation spectra were only resolved through the introduction of *quanta* by Max Planck in 1900. This restricted a previously understood continuous range of radiative energy to tiny discrete packets known as quanta. The

Planck relation e = hv^2 thus forms a fundamental equation in quantum physics and atomic theory.

- The discovery of radiation supplanted the presumed elementality of atoms while providing some harsh lessons in the dangers of experimentation and scientific exploitation. The dawn of the 20th century then saw the rise of a new atomic paradigm (the good old miniature solar system), which was no longer indivisible: all elements regularly undergo spontaneous fusion or fission, and in doing so transform into other elements while absorbing or emitting other subatomic particles or radiation.

Einstein and the Theory of Relativity

The first crack – the constancy of the speed of light (300,000 km/s) regardless of the *reference frame*[3] from which it's observed – was enough to rock the foundations of scientific totalitarianism, which were subsequently shattered with the publication of Einstein's theory of special relativity in 1905.[4]

At the heart of the theory's framework is the idea that all motion is relative, and that physically no reference frame is more or less significant than any other. Here's an analogy: you're working on the railroad tracks and two trains whiz by you, both heading the same direction, side by side, at exactly 80 km/h. But to an amorous young couple, one seated in the first train and the other in the second, gazing distractedly into each other's carriages, neither one nor the other seems to be moving. The idea is then that because all motion is relative, physically neither reference frame (yours nor that of the love birds) is more significant than the other.

If this impartiality is then given a mathematical framework, into which one incorporates the aforementioned constancy of the speed of light (independent from the reference frame from which it's observed), you start getting stranger results.

Time, for example, is no longer absolute but relative, as it dilates as velocity increases: meaning that physical processes (such as the ticking of a watch) slow down, with the effect growing markedly more pronounced approaching the speed of light. On the other hand, lengths contract with velocity. For example, the photo finish of an uncannily fast race between a pair of rulers would show their lengths to have actually contracted compared to that of their stationary counterparts (again with the effect growing markedly pronounced as they approach warp speed).

What's more is that *time dilation* and *length contraction* are forcibly reciprocal, on account of the framework from which they were derived adhering to a complete impartiality regarding reference frames. It's the relative velocity that determines the effect, so if the camera were doing the racing and snapped you standing still at the finish line, your length would appear contracted in the photo. But... which one of you would time slow down for?

This last question, originally proposed by French physicist Paul Langevin, is known as the *Twin Paradox*. The idea is that one twin stays on Earth while the other blasts off in a rocket, travelling at nearly the speed of light, and on account of time dilation reaches the nearest stars after just a few months, then the centre of the galaxy after only 21 years (27,000 light-years away), and the Andromeda Galaxy (1.5 million light-years away) after 28 years. Now, the twin makes a U-turn and returns to Earth in time to celebrate the siblings' joint 56th birthday. The rub is, then, who's the older twin, as there's no preferred reference frame, and to each the other could be seen *speeding away*? The solution involves a practical touch that introduces the third-party perspective: the two twins were originally in the same reference frame, and one left it by accelerating. Both may see the other *speeding away*, but any third-party reference frame can confirm that it was the indeed the astronaut twin who left their original frame of reference:

which gives a physical basis for confirmation without preferring one frame over the other. In conclusion, the twin who stayed behind on Earth would have aged three million years during the trip and been decidedly older.

In that regard, relativity also seems to revel in crushing dreams of space exploration (let alone conquests) and you'll find that science-fiction authors have a marked tendency to set interstellar adventures in a future just distant enough to the extent that the right scientific advances seem implicitly plausible. The issue is actually essential to our evolution and all the more so as, according to the theory of relativity, we can never surpass the speed of light. However, recent theories have started to call this into question.

Besides speed limits, relativity also establishes two other extremely important principles:

- Mass increases with velocity.
- $E = mc^2$, or *rest energy is equal to the product of mass and the speed of light squared*, which you'll find plastered all over the place and is arguably the most recited aspect of relativity.

Einstein later expanded the framework of his special theory of relativity to allow for accelerating reference frames and gravitational phenomena, and the resulting general theory of relativity supplanted Newton's law of gravity as a more complete description of phenomena that identifies mass as a deformation of spacetime. Colossal bodies such as stars, for example, are modelled as spatiotemporal basins that induce the apparent attraction between orbiting bodies and – in extreme conditions – collapse, giving rise to black holes.

The big picture effect of this was that relativity dissolved the long-held belief in the absoluteness of space and time. This all but destroyed the classical paradigms of space and time, and it attacked the very assumptions that went into paradigms of reality as well. Currently, our best understanding is a single, seamless blend of four-dimensional spacetime.

And relativity was but the first earthquake to shock 20th-century physics. It left two pillars standing: *causality* and *energy conservation*, with the subsequent fall of the former constituting a sort of Armageddon in relation to the existing human understanding of the universe – a final blow that would once again be dealt by light itself.

In that era, the invisible hand (notably optics) was interested in dabbling more precisely into the nature of light, and this led to its encounter with yet another anomalous phenomena: the photoelectric effect.[5] The observed behaviour stubbornly evaded the predictions of physical paradigms, that is until Einstein came up with the audacious proposition that light was composed of particles called photons, analogous to Planck's *quanta* and possessing the same energy, hv. This worked, and the credibility of the particle interpretation of light was established through its successful prediction of observed experimental results.

The rub, however, is that Maxwell had already very clearly, successfully and irrevocably demonstrated that light was an *electromagnetic wave*. Einstein justified this by stating that light possesses a dual nature: it is both wave and particle. What's more is that whichever type of behaviour light manifests is subject to the conditions imposed upon it. By this point, time and space are no longer absolutes, and we find a duality in revealed nature (strikingly reminiscent of the duality encountered in our earlier philosophical excursion), which can change at whim according to the given conditions.

Are things starting to look a bit more like poetry than science? But we're not done yet, as now we're just a hop, skip and a jump from asking: does the dual nature of light mean that matter has a dual nature as well?

CONSCIOUSNESS AND REALITY

Louis de Broglie and Matter Waves

A few years later, someone did take that leap: Louis de Broglie, a young eccentric hailing from one of the most prominent families in France, postulated (rather simply) the existence of *matter waves* in 1923, which was experimentally confirmed in 1927.

It was a simple idea: to extrapolate the dual nature of light to everything else. De Broglie accomplished this with wave mechanics and showed that protons, neutrons and electrons (and all matter) have a wave and particle nature and, just like light, manifest one behaviour or the other in response to given circumstances.

As we'll see, the philosophical consequences of these theories, even more profound than those of relativity, deeply alter our habitual concepts of reality and consciousness. With respect to the more rigorous side of the science, we'll simply note that among the pages of mathematics overflowing with black ink, you'll often find the symbol Ψ (Greek *psi*), which denotes the psi function, a mathematical description of the wave nature of the system at hand and a centrepiece of the quantum mechanics and quantum field theory that developed over the following years.

Quantum Mechanics and Quantum Field Theory

By this stage, matter has escaped our intuitive understanding by revealing a wave–particle duality that in the very least upsets the notion that something can't be two things at once. Nevertheless, these are the experimental results, and the conventional interpretation is that both are simultaneously true[6] and that it's the observation itself that evokes one behaviour or another, as well as collapses a superposition of probabilistic states[7] into a single measured outcome that cannot be known or predicted in advance.

This deprives the concept of a particle's position of a certain reality, as the best one can do prior to measurement is to estimate the probability of where it might be. This is also compounded by "commutation relations", such as Heisenberg's uncertainty principle, whereby the determination of certain properties of a particle causes others to be indeterminate. For example, the more precisely one measures a particle's position, the more indeterminate its velocity becomes (and vice versa), whereby it's impossible to know both at the same time. With respect to the beleaguered case of infinitely precise measurement, it also implies that observables need not remain constant when measured repeatedly.

This loss of objectivity precludes the vast majority of classical schemas, and notably both classical and relativistic physics, which generally presume experimental results are determinate and absolute. Indeed, prior to measurement, it's commonplace to perceive a particle as occupying *all* possible probabilistic states, each with its own measurable outcomes. Beyond this, we actually see a *coupling* between the *act of observation* and the *outcome* of the measurement, whereby the former causes a collapse of infinite probabilistic states into a single tangible result.

And all this was followed by the discovery of antimatter to boot, which opens up questions concerning energy conservation and the flow of time. The anti-electron (more commonly known as the positron), for example, can for all extensive purposes be considered a normal electron that experiences time in reverse (i.e., from the future to the past).

We're then called upon to reformulate both energy conservation (both classically and relativistically), as well as causality, which seems to behave intuitively in the macroscopic domain but can get wonky in the quantum (with effects preceding causes, for example).

Layers of Reality

Our instinctive conceptualization of reality, which has already suffered some heavy blows from relativity and wave–particle duality theory, now seems to all but crumble under the weight of quantum scale observations and the discovery of antimatter. In many ways it's more fitting to speak of *layers of reality*, an expression that draws certain analogies with layers of consciousness – which become even more pronounced when one considers that many scientists perceive the conscious mind of the experimenter to act as a participant in the act of quantum measurement ... albeit without precisely defining exactly what the conscious mind actually is.

The conventional interpretation of quantum mechanics is quite clear: systems of particles are best represented as a superposition of probabilistic states that more or less coexist, and it's the rather abstract idea of *observation* that collapses this superposition into a single state with measured outcomes. Observation somehow evokes a particle's properties, which can only be predicted in a probabilistic sense.

These results seem to insist that the conscious mind of the observer acts as a participant (aside from planning and execution) in quantum experimentation, and therefore a new schema is required that accounts for this dynamic, whose exact processes aren't yet known.

Further Entanglement in Quantum Weirdness

As was the case with relativity, the introduction of any strange new theory is rightly met with challenge. Ironically in this instance, we now find Einstein standing on the other side of the fence with his famous EPR paradox

(Einstein–Podolski–Rosen) just before the Second World War. However, recent advances, and notably physicist Alain Aspect's experimental work at the University of Orsay, seem to suggest that quantum mechanics has grounds for dismissing Einstein's claims.

Aspect employed a phenomenon called *atomic cascade* to induce the creation of two "entangled" photons: because they're born simultaneously from the same event, after which they go flying off in opposite directions, certain properties of the photons theoretically must be co-dependent, or entangled, in order to respect certain conservation laws (fundamental laws of nature). When the entangled pair was separated by a distance of roughly 15 metres, Aspect had two separate observers each conduct an identical measurement of one of their properties, namely their spin.[8]

As a quantum observable, spin is naturally indeterminate and prior to measurement exists only as a superposition of probabilistic states. However, the results of Aspect's experiment show that the two spins of the two photons are actually correlated, or entangled, as predicted, whereby the outcome of both experiments is not entirely probabilistic. This is a bit surprising, because the pair of photons behaves as if one of them has immediately learned the results of its fellow's measurement and subsequently rearranged itself to produce a correlating outcome.

The schematic implications are rather profound. Some physicists don't hesitate to speak of a form of "non-separability", whereby physical separation is reduced to a mere illusory construct of our brains. This does away with the anomalous entanglement behaviour mentioned above, as the two photons, which have never actually separated at all, behave as a single whole – like they should. Others speak of *hidden variables*, and some add a *non-reality* to non-separability to spice things up even more.

Beyond Experimental Methodology

Modern physics is characterized by fundamental revisions of our conceptualization of reality via experimental methodology and scientific reasoning on account of phenomena such as:

- relativistic spacetime
- wave–particle duality
- continuous discoveries of newer, smaller and more elementary particles

We see a regular influx of news and discoveries coming from the great particle accelerators in Europe and the United States, so much so that it's often said that modern physics is particle physics.[9] What we once thought was elementary turns out to be just another layer: we have seen how the atom has abdicated its throne to the proton, neutron and electron, which have been subsequently overthrown by the court of even tinier quarks, inside of which a brewing insurrection of undiscovered dinkiness is all but certain.

The concept of elementary particles therefore seems to *want* to fade away, ceding place to a much more evanescent schema of reality that must also account for:

- an *objectivity* that falls apart when observed through the quantum lens;
- a *causality* that can hold its own in the macroscopic domain but gets tossed around in the quantum rink.

Moreover, we have reasonable grounds to conclude that there's a significant shortcoming in the ability of the experimental approach, because its scope of application was not designed to handle things such as the participation of the conscious mind and non-reality. Our options are, then, either to continue using too coarse a sieve or to take the rather unsettling step that accords the conscious mind a role in physics. And there's yet another pill of abstraction to swallow whose side effects are far from innocuous: the role of *information.*

Information has close ties with the abstract concept of entropy, first introduced in 19th-century thermodynamics, which more or less describes the level of disorder in a system. Within a system of molecules, for example, a totally random and chaotic distribution represents a state of high entropy, whereas any order that deviates from this (vis-à-vis position, configuration, etc.) represents a decrease in the system's energy.

The choice to define it this way instead of the other way around is on account of the observation that, in isolated systems, entropy (disorder) always increases: probability, energy transfers and collisions all work toward a dissolution of any established order, not its spontaneous construction.

Formally, this is known as the second law of thermodynamics: the total entropy of an isolated system must increase or remain constant, and this direction toward disorder is known as the "arrow of time". For example, one might define a system under consideration as an entire universe, which is then isolated simply because its all-encompassing size guarantees that it doesn't undergo exchanges with anything outside its borders. The second law then predicts an unflinching path toward an "entropic heat death": a state where disorder is maximized and order can no longer be extracted from the system.

If we instead limit our system to just Earth, the overwhelming abundance of evolving life forms seems to contradict the arrow of time (entropy), as we find levels of order and organization all the way down to those of bodily organs, cells and even biomolecular structures. However, the Earth is notably not a closed system, as incident upon it we have our daily dose of nourishing radiation arriving from the

Sun, inside of which the increasing disorder resulting from the sustained barrage of nuclear fusion results in an overall *net increase* in the entropy of the *isolated* Earth–Sun system.

Thus, without violating the second law of thermodynamics, our autonomic nervous system is able to induce transpiration to maintain a constant body temperature, our immune system can respond to threats, and we can maintain a general homeostatic order – although entropy perhaps has the last laugh if ageing is perceived as an increase in disorder, whereby the relative entropic death encapsulates that of the human.

On that note, in order to move on and discover information's role in all of this, we'll first need to summon a demon.

Maxwell's Demon

James Clerk Maxwell (1831–1879) begins his séance by evoking a molecular system in complete disorder: a room with high and low kinetic energy molecules irreparably intermingled in an entropic death and beyond hope of resuscitation, unless of course ... you cut a deal with the Devil. Maxwell then conjures up a wall separating this system into two halves, and invokes a demonic bouncer to man the only doorway. He whispers into the demon's ear: *Only the high-energy particles get access to the VIP suite on the left. Let the energy-poor filter out to the dive on the right as they approach.* The demon mans the door accordingly, and voilà! The total entropy of the system decreases.

The moral of this tall tale is found in the following reflection: well, just how far-fetched is it to man that door? Is there actually a practical means of decreasing entropy and re-establishing order in a closed system?

Information then comes to light as one of the ways of saying, yes, there is: through synonymizing its collection, retention and loss with a certain *order* that has entropic value, *information = negative entropy* (or *negentropy*). Any practical implementation of the equivalent of Maxwell's demon would then forcibly act as an information processor determining which guests were granted access to which venue, whereby the collection, possession and loss of pertinent information would grant permission for the whole unholy business to proceed without violating the second law, and with chaos a viable route to order.

Information: An Abstract Reality

At this point it's natural to wonder: how does one quantify information in such a way that it can be considered on a par with a measurable quantity such as energy?

Norbert Wiener (1894–1964), one of the pioneers in this area, put it very concisely: *information is information.* That's to say that, like energy, information is something universal and fundamental that has the capacity to take on diverse forms, but whose precise definition is rather elusive. However, the informational perspective seems key, as it's fundamental to a spectrum of approaches situated between the following two extremes:

- **Scientific**: reproducible, objectively measurable results.
- **Unconventional**: where for brevity we'll absorb the pejorative hit and use this term for everything that doesn't make the scientific cut, yet still pertains to the conscious experience and the senses.

Science attempts to grasp information in its raw state, minus the usual transformations and rationale that construct the subjective reality of our everyday experience. In contrast, something more "unconventional" would be, for example, yogic meditative practices involving sensory perception and states of consciousness, where the employment of terms like "ecstasy" to denote out of body experiences has a distinctively fantastical aftertaste, although the affront to modern sensibilities can be lessened with the interpretation of this as an altered state of consciousness.

The approaches on the scientific end of the spectrum are nevertheless encumbered as well, as they are forcibly predicated on the notion that the diverse elements of reality can be separated and effectively studied in isolation. A focus on the whole thus seems to be both outside the scope of what science was built to handle – and exactly what the unconventional side of the spectrum has a knack for tackling.

More toward the middle of the spectrum, recent decades have witnessed neurophysiology use MRIs and EEGs to peer into altered states of consciousness and deep meditation states, identifying notable alpha brainwave patterns. Psychopharmacology has actually conducted research on hallucinogens such as LSD and mescaline to reproduce (successfully) altered physiological states resembling those of ecstatic conscious states.

We can even expand the spectrum to incorporate the often surprising levels of insight pertaining both to the conscious mind and our surroundings that are evinced in great works of art – and it's interesting that the picture painted by quantum mechanics is ready to give the most abstract depictions a run for their money.

While these later phenomena tend to be both difficult to communicate and prove quite fleeting, others include more accessible and more practical examples toward the unconventional end of the spectrum that have been studied. This will be the topic of our present excursion. First on parade: coincidences.

Kammerer's Bowl of Seriality

It actually was a scientist, biologist Paul Kammerer (1880–1926), who first decided to undergo a systematic study of coincidences, which he characterized as sets of events that:

- appear grouped
- occur within a relatively short period of time
- aren't linked causally

Kammerer actually kept a coincidence journal for two decades, in which you will find the following example:

> November 4, 1910. Brother-in-law goes to a concert where he seats in seat number 9. The valet at the coat check also by chance hands him ticket number 9. The following day, brother-in-law goes to another concert where he seats in seat number 21 and by chance receives ticket number 21 from the coat check.

There is also this example concerning Kammerer's wife:

> Wife is in the middle of a novel about a fictional character named Madam Rohan. While on the tramway, she sees a man who looks a lot like our real friend, Prince Joseph Rohan. She believes she overhears someone asking the Prince's doppelganger if he knew whether or not Weissenbach-am-Attersee was a nice place to visit. When she gets off the tramway, she stops by a delicatessen where the clerk tells her that he needs to ship a package to Weissenbach-am-Attersee, and asks if she'd ever heard of it. Later in the evening, Prince Rohan spontaneously stops by our house to visit.

These events are characterized not only by the indisputably small probability of their occurrence in such close proximity, but also by the necessity of subjective observation which relies on the individual observer's interpretation.

Kammerer labelled this type of phenomena *seriality* and saw in it a universal principle of nature that manifested independently of causality. He postulated that a law of seriality, homologous to and as fundamental as the law of gravity, caused similar events to attract through spacetime, and that the coincidences we perceive are rather like the subtle peaks of a looming iceberg to which we've been blinded by an education exclusively based on principles of causality. Through the course of his investigations, Kammerer would touch upon many practical aspects of probability, statistics and actuarial science.

Jung and the Theory of Synchronicity

The Swiss doctor and psychologist Carl Gustav Jung (1875–1961) was both one of the first to recognize Freud's greatness and one of the first to break away from the psychoanalytic movement. He notably introduced, bang on top of Freud's signature unconscious mind of the individual, a collective unconsciousness: a stratification of millennia of human experience expressed through a relatively smaller number of principal archetypes.

In the 1920s, Jung became captivated by an ancient Chinese system of divination known as the I Ching, which was introduced to the West by the pastor and missionary Richard Wilhem. During a 1930s tribute to Wilhem's memory, Jung states:

> The I Ching is based not on the principle of causality, but on a principle that's gone by both unnoticed and unnamed. For the time being, we can refer to it as "synchronicity".

He consecrated a large part of his remaining years to constructing a theoretical framework for this new concept, which culminated in his 1952 work written in collaboration with the Nobel Prize-winning physicist Wolfgang Pauli: *Naturverklarung und Psyche.*[10]

First, it's worth taking a closer look at the I Ching itself, which is a several thousand-year-old system of divination consisting of the interpretation of hexagrams formed by a set of 50 sticks. Its practice, perfected over the centuries and recorded in the *Book of Changes* (which includes philosophical details pertaining to the interpretations), is founded on two hallmarks of Chinese philosophy:

- **Eternal and incessant transformation**: a universe in constant evolution shared between two opposing forces (yin and yang).
- **Imagery**: every occurrence in the visible world is a mere image, or copy, which reflects an extrasensory event (this is reminiscent of Plato's ideals).[11]

This inspired Jung's pursuit of *synchronicity*, which he characterized thus:

- two simultaneous events linked through the senses and not the cause
- two coincident events that aren't causally linked and are in a sense identical or similar
- of an importance equal to that of causality as a principle of explanation

The philosopher Michel Cazenave (1942–2018) elaborates on Jung and Pauli's work:[12]

In the wake of his clinical experience, Jung defined synchronicity on two distinct levels. First, the type of synchronicity that presented itself in his own practice:

phenomena correlated with individual patients, bearing personal signification for individual patients as well as mental conditions correlated with the external factors, events occurring outside the patient's scope of awareness (for example, Swedenborg's famous premonition of the Great Stockholm Fire)[13], [14] and future events. Causality is insufficient to explain any of these phenomena, which therefore necessitates a new schema that acknowledges an objective mental state lying above or below spacetime.

He cites one of Jung's examples that exemplifies both elements of synchronicity and archetype:

One of my patient's wives, a woman in her fifties, had told me that at both her mother and grandmother's funeral, a large flock of birds appeared. I'd heard the story from several others. Later, when her husband's neurosis began clearing up, I noticed for the first time the faint signs of what seemed to be a heart condition. I recommended a specialist who, after examining the patient, wrote to me to say that he didn't notice anything to worry about. However, the patient, however, collapsed on the way home. When he was brought back to his house, dying, his wife seemed already aware: just after her husband had left for the doctor's office, she saw a flock of birds come barrelling toward her house. Naturally, she feared the worst.

Cazenave concludes:

We find ourselves in the face of a series of events that aren't rationally linked but nevertheless possess similarities regarding their physical objective reality and archetype – notably so, given humanity's long history of ornithomancy [...] whose vestiges remain intact today (e.g., albatross, crows, ravens).

He goes on to state:

> If we admit that [...] cases of significant coincidence (which should be distinguished from simple chance occurrences) seem to possess archetypical elements, then one is forced to recognize a temporal correspondence of the physical states with the mentality that identifies the archetype: like an intangible union of two domains that we perceive as distinct in our daily lives.

A more modern terminology might say that *synchronistic phenomena* seem to be characterized by a singular form of spacetime, and in this way spare causality at the expense of suggesting that such a deformation or phenomenon could occur wildly beyond the realms that relativity or quantum theory might predict. Cazenave suggests a new paradigm is needed:

> Synchronistic events are characterized not only by rational relationships, but also by a deeper causality for which spatiotemporal ordering seems to disappear [...]. In the case that two independent physical events give an individual a sense of being able to infer one from the other, as with the case of the woman and the birds, or premonitions, the disappearance of causality is but symptomatic. The current paradigm of physics, which dictates that A engenders B implies that A precedes B, along with the need for a temporal sequence (to whatever extent) to distinguish cause and effect, is negated by synchronistic phenomena, and there's a need to give meaning to the concept of a reality outside of time.

Lastly, Cazenave points out the fundamental role played by interpretation:

It's necessary to note that synchronistic phenomena depend on the interpretation of a certain meaning. The association of death with flocks of birds really only exists for the individual who sees it in this particular light. The term "observer" is also very tenuous, because the woman in Jung's example is not just a simple bystander. The observer is characterized by a unique interpretation that renders them a participant in the event's meaning, which is then relayed from the event to the observer, and back to the event, like a two-way street.

Others can get there too, provided the interpretation is shared.

Jung and Pauli's *synchronicity* is very similar to Kammerer's *seriality*; however, there are notable differences. Seriality deals more with temporal proximity, while synchronicity allows a broader range of interpretation and thereby introduces an additional layer of subjectivity. Seriality also presumes an underlying affinity among correlated events, while in synchronicity these events are drawn together more or less by the gaze of the observer.

In any event, both paradigms are generally characterized by a proximity, a lack of causality, a correlated meaning and a conscious observer; as well as by elements reminiscent of non-separability, entanglement and the role the conscious mind of the observer plays in quantum mechanics. As Pauli exemplifies, many prominent physicists seek explanations outside the strictly scientific end of the spectrum. For our part, we would perhaps do well to notice the slight push toward considering alternate forms of spacetime, and to take the bolder step of linking an interpretation of what this might entail with the role played by information.

Material Witnesses

We've seen the role of the conscious mind come up on the scientific end of the spectrum through quantum mechanics, and more toward the unconventional end of the spectrum with synchronicity (which is notably macroscopic, i.e. visible to the naked eye).

A more pressing question is that if, in either case, we admit the conscious mind is capable of affecting reality, does that imply that the conscious mind itself is actually composed of matter (albeit perhaps a type of matter with which we're not familiar)? Materialistic paradigms of consciousness are, in fact, nothing new and were all the rage with 20th-century neurophysiologists and physicists – which is the subject of our next chapter.

3

LIVING IN A MATERIAL WORLD

Throughout the history of philosophy, religion and lately the evolution of modern physics, reality and consciousness have formed quite the couple, and they'll likely only become increasingly difficult to separate for independent study in the future. Both ends of our schematic spectrum, the "scientific" and the "unconventional", innately implicate the role of the conscious experience, which at this very moment constitutes the central core of the reality around us: it's quite possible that it's the only reality possible.

Consciousness – Way Too Self-Evident?

Despite all that, a precise definition of consciousness seems to slip through our fingers the harder we try to grasp it. Like the terms "reality" or "truth", we're dealing with an abstraction that's so blindingly obvious that it's impossible to draw lines around it, and this is only compounded by additional blurring and ambiguity inherent in traditional philosophical and religious associations of the conscious mind with the spirit.

The scientific end of the spectrum is therefore generally forced to maintain a certain caution on the topic of consciousness, sometimes even refusing to dip in so much as a toe for fear of getting dragged out to sea by an undertow of mysticism.

Many biologists, surfing this past century's waves of positivism, have begun scientifically modelling the conscious mind as an epiphenomenon (a form of by-product) emanating from the cerebral cortex. This schema characterizes the entirety of conscious and intelligent human behaviour as conditioned reflexes. Some even view the cerebral cortex as a blueprint (vis-à-vis learning machines, neural loops, micro-currents and chemical intermediaries) that structurally depicts all aspects of the conscious mind; JP Changeux, author of *l'Homme neuronal* (*Neural Man*), is a well-known member of this community.

Realistic Faults: Colour Perception

In the previous chapter, we saw a significant fault line appear when we took a cursory look at sensory processing and noted that the fundamental reality of sensations such as colour is limited to physical properties such as wavelengths and intensity. Our everyday experiences, which people tend to be in close agreement about, are rather the result of processing through the brain or conscious mind.

As vision is a well-studied phenomenon, we can readily break it down into more detail. First, a beam of light strikes the retina, where three types of cellular cones wait at the interior to capture specific bandwidths (conversely, variations in the functioning of a particular type of cone result in a specific form of colour blindness). It's worth noting too that human retina aren't particularly sensitive on the whole: they have an effective bandwidth of about 380 to 700nm and, for example, are completely oblivious to infrared and ultraviolet spectra.

Light absorbed by the cones undergoes photochemical reactions that we can assume take part in encoding physical parameters (for example, intensity, wavelength and bandwidth), which transform into electrical impulses that travel through a neural network en route to the cerebral cortex, with chemical intermediaries relaying the signal at each synapse. Finally, we can measure the manifestation of electrical activity in the cortex, which is followed a *fraction of a second* later by the individual's perception of the visual stimuli. That fraction seems like an impenetrable wall, however, and here's the rabbit pulled out of the hat:

- *Before*, a wavelength and some electrical activity
- *After*, a sensation

Riddles in the Dark

And before and after *what*, exactly, let alone *how*?

Starting with what we know, that there's an incontestable correlation between physical parameters (e.g., wavelength, intensity) and the *quality* of the sensation (e.g., 700nm wavelengths evoke red, while 450nm are bluer), we can then ask whether sensations can be *quantified*.

Many scientists have conducted research in this area, notably Charles Fabry at the Ecole polytechnique, and the German doctor and physicist Gustav Fechner (1801–1887), who was himself blind for several years. Fechner postulated that sensation could be represented as a mathematical function of the physical quantity it represented, which he approached using a discrete domain of sensory thresholds that has since spawned a Fechnerian school of thought in modern psychophysiology.

However, the approach seems to pass the buck rather than solve the issue at hand, and the results of the larger body of research suggest a resounding *no*: it's not possible to quantify sensations consistently, nor to map them back inversely to the physical quantities from which they originated. The lack of a functional representation of the real world to the senses, as well as inversely from the senses to the real world, in fact not only implies that sensations are not quantifiable, but that there is an actual *discontinuity* – occurring right in between the measurable cerebral activity and its registration in the conscious experience.

This is therefore a phenomenon that lies outside that which science is built to handle, as it doesn't fit within the framework of what can be objectively measured. The study of sensation thus requires a different approach.

From all the riddles, we can ascertain that sensation involves a subtle passage from a physical domain with measurable properties to a *conscious domain* that consequently lies outside the scope of scientific understanding or paradigms; and notably outside reality's dictates pertaining to classical and relativistic spacetime. It's worth noting too that this jump in communication seems vaguely reminiscent of the instantaneous transfer of information seen with quantum entanglement.

On the whole, we can conclude that sensations exist uniquely in the less conventional conscious domain, which seems to communicate by relay with the more scientific domain vis-à-vis the cerebral cortex.

Introspection

The idea that the conscious mind occupies its own domain is nothing new. Without the slightest need for centuries of philosophy, people have always practised introspection that entails the totality of their sensations, intelligence, memory, feelings and emotions. Henri Bergson attempted to encapsulate this with the phrase: *immediate information available to the conscious.* The lack of objectivity identified above suggests that this lies more toward the subjective end of the spectrum and, as Descartes highlighted, is sometimes liable to misinterpretation.

Psychoanalysis tends to understand this through layered schemas, perceiving the conscious mind, for example, as a multi-storey home whose various levels correspond to our conscious experience of the world, ideas and emotions. The basement and foundations represent the unconscious mind: seldom visited and not easily accessible (the foundations in particular) unless you're with Freud or Jung.

Two of Freud's teachers, Pierre Janet (a psychologist, physician and philosopher) and Jean-Martin Charcot (a neurologist), even studied cases of multiple personalities at Salpêtrière. They induced trances and deep hypnotic states in hysterical patients, whereby they observed new and coherent personalities that had nothing to do with the individual's habitual self. These and other studies have led psychiatrists and psychoanalysts to develop the schema of a "normal" individual as a more or less harmonious juxtaposition of a large number of personalities that gave the illusion of a single, unique person. This sort of layering is notably no stranger to philosophy and mysticism as well, which has a penchant for dividing the conscious mind into various states. In instances where symptoms become pathological, harmonious balance is thought to teeter: fragments increasingly take on independent roles, resulting in a lack of coherence.

It's all rather curious, and one cannot help but want to delve deeper into the similarities between the superposition of states in quantum mechanical models and those proposed in psychological schemas of the conscious mind. There are some interesting schools of thought in this area, one being the analogy drawn between the superposition collapse caused by quantum observation and that caused by psychological conditions resulting in the appearance of a single pathological trait; but in-depth studies into the effects of hypnosis and hallucinations are perhaps even more interesting.

Hypnosis and Hallucinations

Hypnosis is often defined as an altered state of consciousness in which the individual is highly suggestible, and in which the senses are malleable to a degree. Hallucinations, which are observed in certain psychoses, involve coherent and realistic experiences outside of reality that are experienced by a single individual (who's often surprised to find that others aren't on the same page).

Peduncular hallucinations, for example, which are associated (but not exclusively so) with legions on the cerebral peduncles, elicit individual sensory experiences. However, the patient is notably usually capable of distinguishing the aspects of their conscious experience that aren't real, which demonstrates a rather unique combination of awareness and affliction.

The 20th century also saw a rise in neurological, biological and psychological research pertaining to the effects of mind-altering drugs. The author Aldous Huxley gave a detailed account of his own experimentation with mescaline in his 1956 work *The Doors of Perception*, which is sort of an early treatise on altered states of consciousness induced by hallucinogens.

Another author, Colin Wilson, not only documented his own experience with LSD in *Access to Inner Worlds* (1983), but also wrote about the neurologist and psychoanalyst Dr John Lilly who systematically experimented with hypnosis, LSD, meditation and sensory deprivation tanks (which he was notably famous for developing). He explain how, in Lilly's 1972 book *The Centre of the Cyclone*, the author reaches results that concur remarkably with what we've identified up to now as the geography of consciousness.

Wilson also cites Lilly's partitioning of the conscious mind and explains that the neurologist would refer to humans as "working biocomputers", likening them to robots that can theoretically spend their entire existence simply responding to exterior stimuli. He says that, according to Lilly, humans occupy an habitual everyday level of consciousness, which has four positive levels above it and four negative levels below it – making nine in total, each of which Lilly claimed to have experienced in the course of daily life through his consumption of LSD. The four positive levels are:

1 concentration and extreme motivation
2 a distinct union with surrounding life forms, which Lilly found through poetry, music and love
3 supernatural powers (telepathy, clairvoyance, out-of-body experiences), which Lilly accessed through LSD
4 Ramakrishna's samādhi: an ecstasy or union with the universal spirit or God

Wilson characterizes the four negative levels as a sort of reflection of their positive counterparts.

As this type of self-experimentation has proven so inherently dangerous and invariably catastrophic, it suffices to say that the effects of certain drugs are characterized by abnormal communications between the cerebral cortex and the conscious mind, with the bypass of normal sensory processing.

Curiously, the sentiment of Lilly's analysis is reflected in many occult and esoteric writings throughout history that promote the idea of a higher self, of which the personality that we know and love would be a mere reflection. Certain occult philosophies also postulate that humans are actually gods that've become so immersed in a game that they've forgotten who they really are – which is also reminiscent of the Platonic notion of a higher reality (ideas) with which communication has been lost on account of an illusory sensory world.

Consciously Constructing Reality

In light of all that we've discussed so far, the study of the conscious mind seems to lie somewhere at the intersection of abstraction, philosophy, psychology, psychiatry, medicine and science, and we find the pieces of a curious portrait beginning to reveal themselves little by little. It concerns the intermediary that *constructs* our experience of the world from the flux of cortical activity induced by physical stimuli.

We've seen quantum mechanics postulate about the conscious mind's interaction with its ambient surroundings, as well as the synthesis of aspects of the conscious experience through hallucination, pathology and meditation. We've also seen superpositions of states arise in schemas of science, psychology, religion and self-experimentation.

Synthesis of all this information is key. In the next chapter, alongside our ongoing investigation of phenomena pertaining to the conscious experience, we'll attempt to walk the walk in this regard and offer up our own schema for dissection. Our construction will involve two very broad, but decisive, assumptions:

- The wide scope of phenomena implicated in the conscious experience would suggest the conscious mind plays an independent role; that's to say that it's not an epiphenomenon or by-product arising from a single physical aspect such as the cerebral cortex.
- The material interactions of the conscious mind, seen in both quantum mechanics and sensory processing, suggest a tangible model; that's to say a *materialistic* schema of the conscious mind.

But what do we mean by matter in this instance?

Consciousness Matters

The concept of a materialistic schema of the conscious mind is not new. In fact, the past 30 years have seen many scientists turn toward materialistic models in the face of a host of perplexing results arising from fields ranging from neurophysiology to psychoanalysis, to cognitive psychology.

The Australian neurophysiologist Sir John Eccles, who won the Nobel Prize in 1963 for his work on the synapse, was one of the first to highlight the contradictions arising from the physiology of consciousness. He published a treatise in 1953 entitled *Neurophysiological Basis of the Mind* that included a very interesting development of a hypothesis he called "the mode of operation of the will on the cerebral cortex":

> Remarkable observations can be made by evoking movements by stimulating the motor areas of the brain in a conscious subject. Though these movements were elicited as normal by messages from the motor cortex, the subject distinguishes quite clearly between them and those that he voluntarily initiated. He will say, "This movement is due to something done to me and not something done by me."[1]

This incoherence between the artificial reproduction and habitual experience then seems to more broadly contradict a materialistic schema that solely considers the nervous system or conditioned reflexes. Eccles therefore suggests that the exercise of one's conscious will produces an effect in the cerebral cortex, and that even a very weak influence involving just a single neuron could cascade into significant cerebral activity.

He then goes a step further and proposes that the cerebral cortex is designed expressly to communicate with the conscious mind, which is otherwise inaccessible through physical instrumentation.

> Thus, the neurophysiological hypothesis is that the "will" modifies the spatiotemporal activity of the neuronal network by exerting spatiotemporal "fields of influence" that become effective through this unique detector function of the active cerebral cortex.[2]

He gives a practical defence for the role assigned to the conscious mind:

> It will be objected that the essence of the hypothesis is that mind produces changes in the matter-energy system of the brain and hence must be itself in that system [...] But this deduction is merely based on the present hypotheses of physics. Since these postulated "mind influences" have not been detected by any existing physical instrument, they have necessarily been neglected in constructing the hypotheses of physics [...] It is at least claimed that the active cerebral cortex conceivably could be a detector of such "influences" even if they existed at an intensity below that detectable by physical instruments.[3]

Numerous other physical paradigms have been proposed, in particular within the various schemas that note spectral tendencies regarding brain function and the cerebral hemispheres (left brain: speech, writing, rational thought; right brain: imagination, spatial interpretation, intuition).

For brevity, in so much as the material nature and environment pertaining to the conscious mind are postulated to lie outside the reach of current observational science, or simply be unknown, we can collectively designate this theoretical body of relevant properties and governing laws (which notably includes possible variation in spatiotemporal structure) as being that of the conscious domain. Following Eccles's example, the detectable physiological manifestations then act rather as a receiver which, conversely, also probes into this unknown region, showing us more or less the refracted end of the white light incident upon the prism.

With our helpful new vocabulary, let's continue our survey of some famous physical models of the conscious mind (although our coverage will necessarily be far from exhaustive).

Other Physical Models of the Conscious Mind

Firsoff's Mindons

First up is the astronomer Axel Firsoff (1910–1981), who gave a rather more fundamental role to what we've coined the *conscious domain*:

> ...the mind is a universal entity or interaction of the same level as electricity or gravity, and we can expect there should be a formula similar to Einstein's famous $E = mc^2$ to exist for it [4]

To draw parallels with Eccles, the cerebral cortex is a noteworthy inclusion among the latter *physical* component. However, Firsoff goes a step further and proposes an atomistic schema for the conscious domain consisting of particles that he calls "mindons", which he describes as being like neutrinos:

> Experimentation into the existence of particles linked to the conscious mind suggests that they lack any gravitational, electromagnetic, or otherwise physical interactions, and in this way resemble neutrinos or

high-speed electrons. Their behaviour would then necessarily be governed by a different set of laws [...] and leads one to conclude that interactions with the conscious mind also follow a particular set of laws and notably its own form of spacetime.

Positrons

In 1967, the British physicist and mathematician Adrian Dobbs proposed a uniquely elaborate atomistic model based on quantum mechanics. He postulated that the conscious mind was made up of collections of particles that he called "positrons", which travelled faster than the speed of light and whose mass was ostensibly an imaginary number. These last two properties are also cited in American physicists G Feinberg and ECG Sudarshan's 1966 postulation of the existence of *tachyons*: theoretical particles explicitly characterized by their superluminal speed.[5]

Interestingly, Dobbs also introduced a conceptualization of two-dimensional time (the first dimension being none other than the familiar timeline of events, while the second is a mathematical abstraction of the probability of these events occurring), as well as a schema for positrons interaction with cortical neurons that is strikingly reminiscent of Eccles:

> We can imagine the brain, or thought itself, as having a series of selective filters designed to eliminate undesired frequencies from incoming bandwidths. Certain signals are then allowed to pass, albeit in modified form, just like an ordinary radio receiver.

Pribram's Holographic Theory

Holography is a lens-less photographic technique that records diffused interference patterns on film.[6] The rub is that if one then illuminates the pattern on this film with a laser, in fact, if any subsection of the pattern is illuminated via laser, you get the appearance of the original object resting in 3D space: holograms.

This phenomenon inspired the esteemed American neurologist Karl Pribram (1919–2015) to develop a schema that likens the reality of our everyday conscious experience to holograms constructed through neurological and cognitive processes:

> Various cells of the brain react to unique frequencies, and the brain then analyses and decomposes these complex patterns, converting them into objects that make up the physical world in a process analogous to the holographic illumination of interference patterns by laser.

Pribram also implicates the Fourier transform, a known mathematical structure used to model physical transformations between frequency and spatial domains, in this cerebral processing.[7]

> ...this question rests with the complementarity inherent in the techniques and apparatus used to make observations. However, in light of Bohr's construction, we can see that complementarity is a fundamental property, both the thing being observed and the observer himself. [...] Fourier's theorem expresses this basic complementarity.

Noting that our 4D spacetime would then be a holographic reconstruction of a postulated *frequency domain*, one can then ask what practical interpretation can be applied to the latter.

> The frequency domain has no relation outside of densities of events: space and time are lost, and their ordinary limits, such as the localization of a given event,

disappear. [...] From a certain point of view, everything happens simultaneously, however one finds oneself interpreting events through various coordinate systems, of which space and time are usually the most useful in our perception of the familiar domain of appearances.

With this last remark, Pribram takes a step further than we've seen up to this point: objective reality is postulated to be a transformation that offers a more readily understood framework in terms of spacetime coordinate systems that fabricate causality and non-localized order. He asks:

Spatial-temporal ordering, non-localized ordering, and all the other orderings, are they entirely a construction of the senses and our brains?

The author Marilyn Ferguson summarizes the whole of Pribram's theory rather succinctly:

Our brain mathematically constructs a "concrete" reality by interpreting frequencies that originate from another dimension – a universe corresponding to the first reality's schema, meaning, which transcends space and time.

In the next chapter, we'll begin construction of our own schema of the conscious mind, which will incorporate many of Pribram's fundamental ideas. This is merely a proposal, our attempt to walk the walk, and to begin to respond to what Pribram concluded from the analogies he noticed with Fechner's psychophysics nearly a century before his time:

There's an obvious need to develop a science founded on the study of the brain that can both embrace modern physics and the intangible nature of human beings.

4

PROPOSING A NEW SCHEMA FOR THE CONSCIOUS MIND

As stated in chapter three, we're going to attempt to synthesize everything we've been studying to show that, rather than shy away from the dark, it's better to take a stab in it. In particular, we'll be choosing to pursue a materialistic schema of the conscious mind. The idea of materializing consciousness isn't new, but nevertheless carries with it the burden of defining precisely what is meant with regard to both matter and the conscious mind – a considerable challenge, given the extent to which attempts to define matter and reality are prone to falling into certain traps, especially in light of the discoveries of physics and research into the senses in the past century.

The schemas we encountered in the previous chapter did not escape this fate, but instead remained consistently vague on certain crucial details. To avoid repeating this, we'll take the more cautious step of building upon recent discoveries that will concretely define the working aspects of our schema within the realm of modern physics.

Trapped Within an Impenetrable Light Barrier

Indeed, we will need to go back over a century in an attempt to free ourselves. Since Einstein, physics hasn't dared venture outside the bounds of relativity, which states very clearly that the *universal speed limit = 300,000 km/s* (i.e. the speed of light in a vacuum). Einstein's 1905 formulation of the theory was itself daring, as it involved an interpretation of the physicist Lorentz's spacetime transformation, to which no one had previously lent sufficient weight on account of its daring. Within this framework, a material body that possesses a mass can approach arbitrarily close to the speed of light – but can never attain it.

Consequently, physicists began to acknowledge the existence of an actual *light barrier*: a wall behind which nothing could be known and is therefore left to speculation – which the physicists themselves compulsively avoided. Only science fiction authors permitted themselves passage, as their astronauts fired up their hypothetical hyper-drives to propel themselves through the barrier at interstellar speeds.

It wasn't until the 1960s that some started to second-guess this dogma and began to seek a means of breaching it that would be consistent within the relativistic framework. Why the 1960s? The era was certainly one of a kind, but the seeds of doubt were perhaps planted a decade earlier with what was then called the *sound barrier*.

The Sound Barrier

Aircraft approaching the speed of sound encounter a number of distinct phenomena pertaining to the compression of atmospheric layers and shockwaves. Early designs that sought to surpass this speed limit often exploded mid-flight, just as if they'd crashed into an actual solid wall. This came to be known as the *sound barrier*.

For many years, it was thought to be impossible to break through it. However, patient research involving wind tunnels, aerodynamics and fluid mechanics elucidated unknown aspects of the phenomena, which led to the design of aircraft capable of breaking the sound barrier. This historical precedent perhaps, then, played a part in inspiring the imaginations of physicists to attempt to pull off a similar feat with the light barrier.

Early Steps

One of the earliest steps was taken by someone we've already met, Louis de Broglie (discoverer of matter waves), who published a reinterpretation of quantum mechanics and wave mechanics in the 1950s, which postulated the existence of hidden variables and a subquantum medium in which particles could travel faster than the speed of light. A late 1950s journal article on physics also referenced a correlation between having a mass that was ostensibly an imaginary number and a particle's ability to travel at superluminal speeds.[1] But it was only between 1960 and 1967 that several American elementary particle physicists (notably G Feinberg, S Sudarshan and C Billaniuk) were able to show, based on relativistic physics, that particles travelling faster than the speed of light could have real-valued energies and momenta, whereby these properties would be observable and measurable.

Feinberg and the Theory of Tachyons

According to relativity, a particle has a *rest energy* given by $E = mc^2$ and the greater its velocity with respect to a particular reference frame, the greater its total energy would be measured to be from said frame. However, the theory's mathematical framework dictates that attaining the speed of light would require infinite energy (which lacks physical meaning), while surpassing it would imply that the particle's energy is an imaginary number (which also lacks physical meaning).

Feinberg had the unique idea of considering the possible existence of particles that *permanently* travelled faster than the speed of light, in which instance the relativistic framework shows that their momentum would be real-valued, and hence physically measurable. According to this hypothesis, there are then three very broad categories of particles:

- **Subluminal**: particles travelling slower than the speed of light (aka *bradyons*, named for the Greek *bradus*, meaning slow). This constitutes all the matter we're familiar with (electrons, protons, neutrons, etc.).
- **Luminal**: particles travelling exactly the speed of light (aka *luxons*, from the Latin *lux*, meaning light). This includes photons (tiny quanta of light that manifest its particle nature).
- **Superluminal**: particles travelling faster than the speed of light.

Feinberg preferred to call this last superluminal group *tachyons* (*tachus* is Greek for rapid) and postulated that they possessed some strange properties. For example, tachyons' energy would decrease with speed (conversely, the pathological case of infinite speed would imply zero energy). The speed of light would also, by definition, constitute a strict *minimal* speed for tachyons.

More generally, this schema identifies two domains (subluminal, superluminal), with the passage betwixt them constituting the third luminal domain, or our aforementioned light barrier. Crossing from the subluminal to the superluminal (or vice versa) is pretty difficult to comprehend. Indeed, as the 1960s progressed, a number of ingenious and sophisticated experiments were designed to detect the existence of tachyons – and all came up empty-handed. Feinberg acquiesced, stating that either tachyons didn't exist, or they were not searching for them in the right spot.

This sort of thing isn't exactly unfamiliar in the world of physics. For example, it took 20 years to detect neutrinos after they were first postulated theoretically. However, aside from this, the existence of tachyons is also a rather loaded question, as the experimental confirmation of superluminal particles would imply the existence of superluminal spacetime, which is generally accompanied by notions of temporal inversion and a loss of causality.

Feinberg illustrates this with an example. In our normal *subluminal domain*, we can observe an electron and positron (or anti-electron) annihilate, whereby their disappearance results in the spontaneous appearance of two photons having a total energy equal to that of the particles they replaced. One then observes a sort of death (electron–positron annihilation) prior to a birth (two photons). In the *superluminal domain*, the order of the phenomena would be inverted, hence birth before death. It's like watching a film in reverse, with future events preceding past events, which is a categorical violation of causality: this is one of the reasons that many physicists doubt that tachyons can even exist.

Super Relativity

In light of these propositions, in 1972 Régis Dutheil began to attempt to extrapolate a version of the theory of relativity that would encompass superluminal particles. Interestingly, a mathematical analysis of the theory's structure seems to suggest that there are two, and only two, homologous ways to construct Einstein's theory of special relativity: a *subluminal* and a *superluminal*. The theoretical structure of the resulting superluminal domain would be quite different, although in many ways symmetric, to our own subluminal experience – in particular with regard to spacetime.[2]

For the purposes of our proposed schema, we've now come across a crucial ingredient, as to construct our materialistic schema of the conscious mind we need to detail:

- a *type* of matter to assign this role;
- a conscious domain that begins to define the properties, nature and interactions of said matter.

Rather than reinvent the wheel, we can take advantage of the existing framework for this relatively unexplored form of matter and explore what comes of applying a schema that assumes:

- The *conscious mind* is composed of *superluminal matter.*
- The *conscious domain* is the *superluminal domain.*

We can then simply explore if and how well this schema performs when asked to explain the many diverse phenomena we've encountered in the previous chapters:

- Why does the conscious mind play a role in quantum mechanics?
- What is the mechanism behind the apparent non-separation observed in quantum entanglement?
- What are the missing details regarding sensory processes?
- Is there any supporting evidence for certain "unconventional" approaches (for example, Jung's synchronicity), which can sometimes have a ring of truth to them?

The World in a Light Cone

To better understand our new superluminal schema, it helps to take a cursory look at a particular graphical representation commonly employed in relativistic physics: the light cone. For visualization purposes, we will restrict ourselves spatially to a 2D xy-plane and consider the normal z axis running perpendicularly through this instead to represent *time*. At the origin, we imagine a source of light that spreads out in the xy directions as *time* increases, thus forming our light cone.[3]

The interior of the light cone represents everything that we can observe without going faster than the speed of light. The entirety of the observable subluminal domain is therefore represented by spacetime coordinates on the interior of this cone. If one draws a line between the coordinates of different events – the sequence of events that make up an individual's life, for example – the resulting path is known as a "world line", which describes the unique path of a person or object through spacetime.

Indeed, the abstraction in this graphical approach tends to invite the question as to whether time *must* be just like this; or if it is rather like a handy projection of coordinates that facilitates our experience by adding a gradual illusory flow, whereby we progress through the passing years. Experimentally, this is supported by the fact that we can't measure time directly: our actual physical measurements pertaining to time all depend on the observation of spatial phenomena – anything that's *regular enough* gives us a clock whose hands then tells us how much time we've experienced.

In any case, as our schematic decisions have resulted in the inheritance of relativistic properties, the paths of world lines in the subluminal domain within the light cone are subject to a few additional restrictions:

- The speed of light cannot be exceeded (graphically, the angle of the world line be shallower than that of the light cone).
- One cannot jump into the future.
- One cannot travel into the past.

Our Superluminal Domain Ran Out Of Time

However, everything travelling faster than the speed of light – that's to say the entire superluminal domain – is represented by the regions outside the light cone. This domain no longer has a speed limit, but rather strictly enforces a speed minimum (if this were a highway, it'd no longer be possible to go *below* 85mph, and no one would bat an eyelid if you felt like doing 200mph or even 1,000mph, etc.).

At the same time, we can list three noteworthy overall effects pertaining to time that seem to share a common theme:

- According to relativity, as one approaches the speed of light, time slows down. Attaining the speed of light is then tantamount to time slowing so much that it doesn't pass at all.
- As one's velocity goes beyond that, toward infinity, the graphical representation of the world line unravels closer and closer to the *spatial xy* plane, and the time component becomes entirely *spatial*.
- Relativity's unification of spacetime, and certain resulting abstractions (as seen in the light cone models), suggest that time rather behaves as a coordinate projection on par with spatial dimensions.

At the very least, the implications of the two limiting cases warrant investigation in their own right, but we can take a leap further and, grasping at the overall trend, choose to associate our schema of the conscious domain with superluminal domain, but in a way that removes all temporal properties from its spacetime: such that there is no flow of time, and all events have a complete instantaneity.

This audacious claim is not without certain philosophical and religious precedents. Buddhism, for example, embraces the idea that past, present and future are nothing more than illusions, and that the path to enlightenment leads to an instantaneous interpretation of all events.

However, more immediately, we have to respond to the need to more or less define the now missing time component of our schema's structure in practical terms. To that end, we'll again attempt to build upon the above observations and postulate that time is entirely absorbed into spatial dimensions: that is, our *superluminal spacetime* will be entirely *spatial*.

Three Universes Instead of One!

In summary, our schema postulates three domains, each of which experiences its own unique form of spacetime:

- **The subluminal domain**: our everyday experience of matter travelling less than the speed of light.
- **The luminal domain**: that which travels at the speed of light (notably photons), graphically portrayed by the light barrier (some perceive the time experienced by a photon as simply zero, lacking both spatial and temporal components).
- **The conscious domain**: our customized version of the superluminal domain (the hypothetical home of tachyons) with a purely spatial spacetime.

The luminal domain acts as an interface or even like a mirror in some ways between its two counterparts. By way of analogy, we can relate our personal experience in this schema to that of somebody's home. We're very familiar with what goes on in the interior (i.e., the subluminal). The exterior can be quite wild and fantastic (the superluminal). In between we have these walls (the luminal), which form a light barrier with their own interior domain.

The physicist Costa de Beauregard expressed a similar sentiment in *Le second principe de la science et du temps* ("The second principle of science and time"):

The framework of the material cosmos is not so tightly woven as to render it self-sufficient and, all things considered, is rather an interior inner than an exterior.

Repurposing Causality

Aligning the conscious domain with the superluminal gives us immediate recourse to propose solutions to the EPR paradox (which seeks evidence of superluminal information transfer), as well as a place to relegate phenomena that seem to elude observation (such as the aforementioned hiatus between electrical activity in the cerebral cortex and the actual experience of sensation). However, the choice to remove all temporal properties and render this spacetime completely spatial warrants some preliminary clarification: what becomes of *causality*, for example?

The essence of our schema is that, as relativity suggests that time and space are on the same footing, time need not be accorded a special precedence and spacetime can be rendered completely spatial (as several limiting cases suggest). Consequently, causality could be re-associated with a spatial dimension, but its temporal correlations would certainly disappear, which one can visualize as now purely spatial world lines.

However, as we discussed above, time in our everyday subluminal domain is also associated with another phenomenon: *entropy's arrow* (aka the arrow of time), or the second law of thermodynamics, which states that entropy (disorder) must constantly increase. Interestingly enough, certain theories suggest that the effect in the superluminal domain must be the inverse: order and information (negentropy) must constantly be on the rise, which for brevity we can refer to as *negentropy's arrow*.

With our schema's spatial dimensions being robbed of all temporal correlations, negentropy's arrow then seems like a likely candidate to replace causality. Indeed, in our everyday concept of time, causality serves as a principal method to organize our perception of information (e.g., *before*, *after*), so the substitution seems analogous. We can also note in passing that replacing causality with negentropy's arrow yields a schema that is much more able to explain synchronistic phenomena (correlated events that lack causal relations).

By this stage, we've established some reasonable guidelines for defining the inner workings of our proposed conscious domain, but it's now time to return to our main goal, a materialistic schema of the conscious mind, and address a fundamental missing piece of the puzzle: how can communication exist between this conscious domain of the superluminal and our everyday subluminal reality?

Establishing Comms

To recap, the flow chart of our proposed schema of the conscious mind has more or less followed a path of least resistance, based on choices relating to and interpretations of existing results:

Proposed Schema of the Conscious Mind
- *materialistic*
- conscious mind is independent phenomenon
  - *subluminal domain*
    - everything travelling less than the speed of light
    - our everyday experience
    - entropy's arrow, aka the arrow of time (increasing disorder)
    - causality
  - *luminal domain*
    - everything travelling at the speed of light
    - photons
  - *conscious domain (superluminal)*
    - everything travelling faster than the speed of light
    - tachyons
    - purely *spatial* spacetime (instantaneity of all events)
    - negentropy's arrow (increasing order, information)

This was fundamentally based, a priori, on the widely held interpretation of quantum mechanics involving the participation of the conscious mind. Although this phenomenon is exclusive to the microscopic domain, our schema has yet to establish *any* of the mandated communication between the conscious and subluminal domains, let alone hint at how an individual might thus divine the nature of what our schema postulates would be a universe apart.

To continue to grope around in the dark, we'll stick to our signature methods and this time borrow from Eccles, namely his interpretation of the brain as a sort of receiver to which the conscious mind has the ability to transmit signals. Eccles supposed that these signals were too weak to lie within reach of modern scientific instrumentation.

At this point, we have enough to start filling in the blanks as we go and can allow ourselves to speculate a little as to how this would all actually play out. Our conscious domain seems to be characterized by two fundamental properties: instantaneity and increasing information. This then begs the question: why does the totality of this seem to be left out of what's detected by the cerebral cortex? Actually, if this weren't the case, then the subluminal universe that we live in wouldn't exist, because it would depend on the filtration of this information into causal sequences that engender an apparent flow of time. The cerebral cortex acts as a computer receiving this information, from which it constructs our entire perception of the subluminal universe.

Indeed, the importance of filtration has also been notably pointed out by the Nobel Prize-winning philosopher Henri Bergson, and is closely linked to biological receptors. The very act of filtration, along with the inherent instantaneity, information and order, suggests that the conscious mind occupies an infinitely richer state in the conscious domain.

In this respect, the world line representing the path of a human life, from birth to death, in subluminal spacetime (as a sequence of causal events) would have an exact homologue in the superluminal universe. But the arrangement of this purely spatial world line would be neither causal nor temporal in the usual sense of the word. It is no doubt negentropy's arrow (order and information) that presides over this assembly, whereby the distribution of information follows laws of affinity and signification which, although stationary from our point of view, the superluminal domain is in a constant state of establishing.

Comparison and Extrapolation

We have reached the point where we can compare certain aspects of our schema to analogous proposals that we've seen set forth by Firsoff, Dobb and Pribram.

Firsoff atomized the conscious mind into neutrino-like particles called mindons, which he postulated were capable of interacting with cortical neurons. Our schemas become very similar in the event that neutrinos can penetrate the light barrier (perhaps becoming subluminal antineutrinos) and transmit information to cortical neurons.

This is also the case with Dobbs, who proposed a quantum mechanical schema explicitly founded on the existence of superluminal particles that he called "positrons" and who also postulated the existence of a unique, two-dimensional form of time (one the familiar timeline, the other virtual and probabilistic). Fundamentally, there is no overt contradiction here with our interpretation. It's simply another way, perhaps a bit more classically flavoured, of elucidating these phenomena, and we can't say that the two schemas are mutually exclusive.

However, the analogies with Pribram's schema, and also that of physicist David Bohm, are perhaps the strongest: the characterization of our everyday perception as a 4D hologram whose spatiotemporal coordinate system is the work of a cortical computer is embraced all around.

In fact, Pribram also cites the experimentally demonstrated reception by the cerebral cortex of spatial frequencies of a non-temporal nature (produced in the lab with various diffraction gratings). Initially, David Hubel and Torsten Wiesel (winners of the 1981 Nobel Prize for Physiology or Medicine) discovered that numerous visual cells in the cortex correlate with specific orientations. Over a decade later, many laboratories used diffraction gratings to demonstrate a selective reception of spatial frequencies in various lobes of the cortex.

In fact, we can likely explain many aspects of our own model as representing a special case of Pribram's model as applied to superluminal particles, which wave mechanics suggest would possess a subluminal phase. If we take the frequency to be purely spatial, we get a means of interaction with specific cells in the cerebral cortex, which would then reproduce the subluminal holograms that constitute our normal perception of reality: to each subluminal particle a superluminal wave that's nothing more than a complimentary aspect of reality. The beats of the interference patterns between multiple waves would actually travel slower than the speed of light and carry energy, which implies that it's rather the interference which the brain observes and renders into a localized subluminal particle (Pribram's holograms).

We also find some additional similarities in Pribram's implication of not only the Fourier Transform, but the Inverse Fourier Transform as well, in the cortex's processing of this information between the spatial and frequency domains:

> The response to this question (is the brain capable of functioning according to inverse transformation) depends on the capability of the Fourier Transform and its corollaries to transform ordinary spacetime, where immediate causality reigns, into an involuntary order, non-locally distributed, in which only correlations exist. This capability is referred to in informatics and statistics as Fast Fourier Transforms, each time, and at whatever level, these correlations should be calculated. It's at the foundation of CAT and PET scans that form limited images in correlation with this domain of transformations.[4]

Endowing the cerebral cortex with the ability, on rare occasions, to invert the perceptional transformation of information would, for example, provide a means of accessing acausal correlations pertaining to synchronistic phenomena. This would also corroborate the viability of chapter two's entire spectrum of schematic approaches: between the *scientific* (localized, causal) and the un-conventional (non-localized, acausal, also seen in quantum entanglement).

Evidently, things should be the opposite in the superluminal conscious domain: superluminal particles would be associated with the properties and beats of subluminal interference patterns. The interference patterns themselves would travel faster than the speed of light. The simultaneity of superluminal spacetime would render wave–particle duality no longer complementary but unified, such that the superluminal observer would observe, at the same time and without contradiction, a superluminal particle that's simultaneously a wave and a particle (that's to say, something that's simultaneously *localized* and *non-localized*).

The superluminal would therefore benefit from a much larger body of information, making it more fundamental and complete than its subluminal counterpart. Our universe would simply be a holographic projection, a subluminal reflection, a grand reduction of information subject to an inevitable entropic degradation with the flow of time.

Philosophical Implications

Every biologist, mathematician and physicist who proposes a schema that pertains to the conscious mind is called upon afterwards to present their corollary model of the universe, from which one can then derive meaning concerning abstract concepts such as finality, freedom or determinism. Pribram was no exception to this rule, and he himself subscribed to a Platonic conception.

Our reflections on the conscious mind have also driven us to formulate responses to certain fundamental philosophical questions, the first of which is causality. In our daily subluminal universe, destiny is governed by two principles: causality (cause precedes effect) and the flow of time from which it is dissoluble. Each human is destined to follow a series of causal sequences that unravel from past to future along their world line.

What new interpretation of causality can we then offer in light of our new schema?

These same events and causal sequences find themselves newly arranged on a world line in the superluminal spacetime of the conscious domain according to laws of affinities and significance (assimilated from synchronicity). The non-temporal spacetime affords an instantaneous and acausal perception of all events that is constantly evolving toward a state of maximum information. The cerebral cortex finds itself at the interface of two universes, capturing this information and projecting it into causal sequences, which are in reality arbitrary, for the benefit of reconstructing a subluminal spacetime that's more amenable to human perception.

Then what about destiny versus free will?

This issue has occupied philosophers throughout history. Some are convinced of an unwavering destiny, such as the relativistic physicist Arthur Eddington, who said: "Events don't happen, they're there and we find them on the path we're on." Others learn toward an absolute liberty, and in more recent years we've seen versions that accord more limited styles of freedom to certain events. Philosophical opinions around this question have a tendency to mimic choices between idealism or materialism, with the former preferring determinism and destiny while the latter opts for free will. But there are exceptions. Democritus, for example, might be considered both materialist and determinist.

Our own schema would seem to offer a new perspective, in that it tries to reconcile materialism versus idealism and destiny versus free will. Rather than perceive these as opposites, they can be absolved into two complementary aspects of the same reality, whereby we have our new principle of complementarity: destiny belongs to subluminal spacetime, and free will to the superluminal.

From the moment of a human being's conception and birth, they exist and are projected on to a subluminal world line that represents their destiny (the events of their life in causal order). This is deterministic. But when this world line encounters certain series of causal events, there is the possibility of exchanging one sequence of events for another: which occurs on the level of the superluminal conscience that perceives events according to information. This is where free will exists.

In the superluminal spacetime, this happens instantaneously. That's to say that from a strictly subluminal point of view, it would be impossible to notice the substitution of one sequence of causal events for another. Everything happens as if the original sequence never occurred and maintains a strict subluminal perspective of a unique world line whose destiny is *ne varietur*.

Even a time traveller exploring the future on their own world line would see immutable events and be forced to conclude an absolute determinism. But if this voyager managed to penetrate the superluminal spacetime of the conscious mind, they would notice that this immutable character is but an illusion. Ultimately, it seems that the antagonistic relationship between destiny and free will only manifests through the lens of a single spacetime.

Finally, such philosophical musings always lead to another question:

What becomes of the conscious mind after death: does it disappear, or does it continue to exist?

This question, which is perhaps even more important than that regarding of the nature of consciousness, will be the subject of the next chapters.[56]

Attempts to Break the Light Barrier

In September 1993, Enders and Nimtz of the Institute of Physics at the University of Cologne, France, published an article in the *Journal de physique* entitled "On Superluminal Barrier Traversal". Their experiments employed wave guides to orient millimetre–centimetre radio waves and prevent their diffusion. The authors observed the lateral propagation of wave packets, or photons, travelling faster than the speed of light, whose thickness they measured to be several centimetres.

In the United States, the physicist Torsten Alvager of the University of Indiana studied the propagation of light through fibre optics and observed evanescent photons travelling faster than the speed of light. This phenomenon can be likened to the experiments of Enders and Nimtz.

PART TWO

5

A BRIEF HISTORY OF DEATH

At the dawn of time, death was baptized in a pool of fear and respect, and very quickly beliefs surrounding the fate of the deceased quickly spawned supernatural entities whose good graces it was thought wise to solicit.

Prehistoric Burial Sites

Remnants of the most ancient funeral practices date back to the Middle Palaeolithic (100,000 to 35,000 years ago). Eight Neanderthal skeletons were discovered in Shanidar, Iraq, laid to rest on beds of flowers. Another tomb discovered in Qafzeh, Israel, revealed the remains of a child placed next to a fallow deer, with the child's chest covered with burned ostrich eggshells and with a large block of limestone placed nearby. Even Neanderthals, one of the few remaining representatives of our primitive ancestors, appear susceptible to the need to ensure a measure of continuity amidst the abruptness of the transition between life and death, and to the sentiment that *something* persists beyond the unseen.

When *Homo sapiens* hit the Upper Palaeolithic Period (35,000 to 10,000 years ago), there was a proliferation of funeral rites and a marked development in burial tombs. Large stones slabs were discovered protecting a woman's skeleton in Saint-Germain-la-Rivière (Gironde). In Sunghir, Russia, ochre pigment covers the remains of a man, along with the bottom of his tomb and burial possessions: bracelets, ivory (mammoth) beads, and flint tools. Traces of carbonization on the beads suggest the funeral involved a form of ritualistic incineration.

Cemeteries seem to have waited almost until the Mesolithic Period (10,000 years ago) and, curiously, these early villages of the dead actually predate those of the living, for which the earliest remnants date back to the Neolithic Period (4,000 years ago). Red ochre was used at the burial sites, while the chambers were generally full of flint tools, perforated shells and teeth. The tombs themselves are covered with stone slabs or stacks of stones under which, in many cases, lies a small hearth containing an offering of deer or boar.

With the immobilization of human populations during the Neolithic periods, the dead were habitually buried near where they lived, sometimes even inside it and accompanied by burial possessions such as tools and jewellery. However, nothing survives from this early dawn of the modern era that would suggest a special arrangement of tombs or specific burial rites.

Then, beginning with the prehistoric Chasséens civilization (2500 to 1500 years ago) in France, a new type of burial ground appears to take hold: dolmens, menhirs and megalithic structures built from large stones that marked ancient landscapes the world over.

One would think a new form of religion had gripped the prehistoric world, inciting the erection of solid enclosures which, in many cases, continued to amass the dead of the surrounding growing populations. Entire families have been found within four-sided structures formed by slabs and roofs of stone. Passageways made from slabs or packed earth sometimes led to funeral chambers. Knives, axes, crude vases and clothing were placed inside, and sometimes sculptures and engravings at the entrance seemed to evoke a distinctly female form (with breasts, a contoured nose and often a necklace) reminiscent of a protectress, a venerated goddess, an eternal mother: traces of the first religion.

From a certain vantage point, these burial rites have perpetuated themselves into the present, as our modern age has no shortage of individual or family burial sites, tombstones and urns, all of which seems to convey the idea that our deceased have the right to a final resting place.

Ancient Visions of the Afterlife

Through paintings and the written word, our first glimpse into the great beyond was passed down through the Ancient Egyptian's depictions of the afterlife. Their cult of the dead originated with the god Osiris, who, legend has it, was assassinated by his jealous brother Seth. His body was cut up into pieces and tossed into the Nile. However, the goddess Isis hunted down the pieces of her butchered husband, which were then bound with bands of cloth. The ritual reanimated Osiris, thereby paving the way for mummification, and he and Isis later conceived Horus.

On Earth, mummification is explicitly identified with attaining the afterlife and immortality. After embalming, mummies are enclosed in a sarcophagus and buried in a necropolis that honours a particular goddess and faces the west. The goddess welcomes the deceased, who then departs in the direction of the setting Sun equipped with burial possessions such as the *Book of the Dead*, tools, personal effects and even – for the wealthier – figurines representing servants. The road is perilous, fraught with pitfalls that the *Book of the Dead* enables one to evade.

The souls who make it through are met by a grand tribunal presided over by Osiris, where the goddess Ammit, the great "devourer of the dead", weighs each heart against a single feather and consumes the hearts that are weighed down by ill deeds, thereby trapping the deceased. Those who make it to the Kingdom of Osiris live out their days in peace, among fields of reeds and agriculture.

The Egyptians envisioned death as a continuation, the beginning of a new life that in many ways resembled life on Earth. Each individual was thought to contain six distinct elements: three material (body, name and shadow) and three spiritual (spirit, mind and *ka* – a mysterious principle that protected the deceased during their voyage).

In the grand scheme of things, Western civilization has seen very little variation on this theme over the past 5,000 years. If Mesopotamians tended toward the despair of a hopelessly eternal underworld, reminiscent of that abstracted deprivation of God known as the *Sheol* in Hebrew, the Greeks and Romans preferred to paint the afterlife with an excess of anthropomorphism entailing very human forms of recompense and punishment.

To the Greeks, death was both fearful and a critical rite of passage, as illustrated in the *Iliad* and the *Odyssey's* depiction of Thanatos (death), the brother of Hypnos (sleep), with life in the underworld being terrestrial-like but decidedly robbed of joy. They even gave it a geography, pinpointing the entrance on the map. Souls pay the ferryman Charon an obol for passage over the River Styx, after which comes Erebus, or darkness, where the ferocious three-headed dog Cerberus stands guard over shadows of those wandering souls who didn't receive proper burial rites. Like the Egyptian legends, the spirit is weighed before Hades, god and ruler of the underworld, and three judges: Minos, Rhadamanthus and Aeacus. The outcome is either eternal damnation in a lake of boiling pitch, or eternal youth among the Elysian Fields, where the River Lethe (the river of forgetfulness) waters the final resting place of victorious souls, while more mediocre souls wander in meadows of asphodel.

As you can see, the afterlife of the Ancient Greeks isn't all that different from the modern Western notions of heaven and hell. Their burial rites, which were quite explicit (washing of the body, bands on the eyes and on the chin, the laying of the deceased on a high bed, mourning, coffins made of cypress, burial chambers), clearly set a precedent for those of the Middles Ages right up to the modern day. There was even a version of Halloween: in Greece this took place on two days during the festival of Anthesteria in February or March; in Rome, it was known as Rosalia, where members of the deceased's family would gather around their tomb and cover it with rose petals. The Romans, for that matter, rallied around the Greek vision of the beyond and consequently changed very little.

We therefore have two populations for whom the afterlife was more or less an agreeable, or disagreeable, amplification of terrestrial life. A few philosophers, however, delved a bit further into the abstraction and came out with rather more.

Abstraction and the First Theories of Reincarnation

The mathematician whose household name seems to span millennia, Pythagoras, is ironically one of the first Western philosophers on record to have demonstrated an interest in reincarnation and metempsychosis – the transmigration of the soul.

Plato, who in many ways was his heir, inadvertently took several steps down the same path by schematizing a Realm of Ideas from which terrestrial life represented a banishment whose only escape was death. In this obviously very personal vision, which demonstrated a lot of gall in dissenting from its contemporaries, he also considered each new incarnation as the effective punishment for poor choices, which notably carries the sentiment of ancient Hindu teachings dating back to 4,000 BCE. However, Plato didn't modify a vision of the afterlife (perhaps to avoid too much shock and awe), and in the Myth of Er in Book 10 of *The Republic*, we find descriptions of heaven and hell that conform to those of Greek mythology.

Whereas Aristotle equated the disappearance of the body with that of the soul, Plutarch (46–119 CE) discerned three distinct parts – body, soul, and intellect – and postulated that only the soul experienced reincarnation, through a punitive process similar to Plato's.

In comparison, certain Celtic and Gaelic conceptions of the afterlife appear distinctly original. According to the Druids, those souls whom death has liberated embark on a path of interplanetary and interstellar travel that's predestined end is fusion with the Sun. The universe itself is divided into three circles: the *Cuegant* (the realm of divinity, the domain of the absolute and of the inconceivable), the *Gwynfyd* (bliss, a sort paradise for more advanced souls), and *Abred* (the domain of humans fighting for their own ascension). The Gauls were, then, reincarnationists and understood death as a simple movement toward the afterlife, a place where time is abolished.

However, all of these variations seemed to lack a certain foothold and remained at the outskirts of the monochromatic panorama of Mediterranean religions that dominated the Western world for centuries – and to which we are in fact heirs, just as we were with Latin, the legal system and aqueducts.

The Old and New World

Since the High Middle Ages, after the fall of the Roman Empire, two great religious conceptions of life and death dominated the Old World. The West saw a proliferation of Judeo-Christian religions (such as Islam, Judaism, Catholicism and Protestantism): monotheisms grafted on to the trunk of their polytheistic ancestors. Meanwhile, in the East, Hinduism, Taoism and Confucianism had been evolving for millennia.

As for the New World, the Incas and Mayas practised burial rights that closely resembled those of the Egyptians, even employing their own *Book of the Dead*. And while the many religions of civilizations native to the continental Americas and Africa (which frequently embraced the idea of soul transmigration) are equally fascinating and distinct, a proper study that avoids deformation and reduction through the influence of 16th- and 17th-century Catholic and Protestant invaders is beyond the scope of the current text.[1]

Turning to Face East

It would seem that the historical tension between Eastern and Western cultures continues into the afterlife. In Buddhism, Hinduism and Confucianism, death is perceived more or less as an inadvertent part of an eternal cycle that defines human destiny.

Buddhism perceives this principally as one of the many effects of ignorance and desire. Birth and death are actually mutually consequential, marked by pain and inscribed into the never-ending circle of transmigrations: the Samsara. Everything that is born is called upon to age, to die and to be reborn.

Where Hinduism tends to embrace the subsistence of a permanent entity after death, which can assimilate the Western concept of the soul and which reincarnates until reaching a state of moral perfection, Buddhism, on the other hand, prefers to consider a more fleeting flux of desires and actions that lead to a reincarnation. The goal is then to exit this trap and to reject both birth and death, the latter being nothing more than the result of the desire to live and the imperfection engendered by rebirth. This also robs death of any finality, because it's immediately followed by a new existence set in motion by one's good and bad deeds. In fact, in Buddhism death doesn't exist at all, and the role of an ultimate finality is replaced by the sought-after end of one's cycle of reincarnations. Sufferance and separation are relegated to a domain apart, to which the self that is born, desires, suffers, ages and dies cannot belong – because it is immortal and knows no pain. Thus, we find death is excluded by both its transient nature and its dissociation from what is considered to be the true and immutable self.

Through China's vast landscape, currents of Buddhism, Taoism and Confucianism flow and merge behind various interpretations of reincarnation. Taoism, like Buddhism, illuminates a place beyond the reaches of death, outside both space and time, where there's no longer a past or present and the soul is neither alive nor dead.

The Tibetan Book of the Dead, or *Bardo Thödol*, predicts that one's death is preceded by a subtle noise, a whistle or a rumble, after which the deceased becomes immersed in a great luminous fog and is called upon to leave their body. Dead loved ones arrive and are heard, but the freshly deceased soul cannot respond. It integrates into a new body, one that can pass through material objects and immediately travel at great distance, and which is bathed in a blinding light. Only the souls that were sufficiently trained in spiritual practices while alive are able to discern the path that will deliver them from the cycle of rebirth. Those who are lost must face a new realm of terrifyingly violent apparitions comprised of their own introspection. If they survive, there is a chance of deliverance, but those who succumb descend into a vortex of reincarnation where but a single hope remains: to resist the temptation to reincarnate.

The Eastern tendency is then to face death with almost a negation, a banalization which the adept turn to spiritual practices to tackle.

Sunsets in the West

The West, on the other hand, with its largely Judeo-Christian influence, seems to prefer an empowering concept of death, feeding a gripping anxiety that places the phenomenon at the crossroads of an irremediable breaking point in one's existence and identity. The afterlife appears to envelop its earthly counterpart, which it seems to amplify in one direction or another.

According to Islam, God interrogates souls individually, weighing their deeds and granting admission to either a garden of pleasure or to hell or torment, where the exact comeuppance is doled out accordingly (flames, shade, greenery, good or bad company, delicacies). The distinction between the two sexes is notably transcendental as women are subject to an unflinching fidelity to their former spouses, whereas men are granted pungent promiscuity.

However, the Church has no right to throw stones. Catholicism and Protestantism describe a similar theme of resurrection at the Last Judgement, whereby a soul is either bathed in heavenly light or condemned to hell for all eternity, although this belief seems to have involved a degree of methodical muting or censorship of dissenting opinions or texts, some of which have a decidedly Eastern flavour (such as biblical passages that speak of reincarnation and the condemning of the theologian Origen's teachings on the pre-existence and incarnation of spirts). In any event, the concept has certainly evolved.

In early Christianity, the dead were believed to fall into a deep slumber where they awaited resurrection and the Last Judgement. The judgement of the early martyrs and saints therefore wasn't immediate, but as their image faded, this form of resurrection became superseded by an immediate judgement of the individual's soul. All the while, depictions of heaven and hell continued to unfurl with ever more clarity.

The Middle Ages saw hell turned into a torture museum: cauldrons, blazing infernos, gallows, scalding baths, iron maidens, vile sustenance. The existence of Purgatory was officially accepted by the Church in 1254, allowing souls to remain in a transitory state where further purification is possible prior to the corporal resurrection that will decide their eternal fate. This was an important point of contention between Catholics and Protestants, with the former preferring to open additional access into paradise (perhaps to soothe anxiety among the souls of the living), and the latter adhering austerely to the antecedent. Be that as it may, the idea was then assimilated into the concept of limbo, which was later viewed as a lower registry of hell. Paradise becomes a garden of delicacies, an eternal meadow that souls enter via a long tunnel or a ladder (according to the vision of St Perpetua). This development was rather conspicuously burdened with a certain level of anxiety that, along with the resurrection itself, seems to foster morbid fantasies (such as skeletons, dances macabre) and the construction of ever more elaborate tombs.

The High Middle Ages were an epoch of intense faith, but still welcomed death as a natural phenomenon. Philippe Ariès (1914–1984) used the word "tame", observing a sort of indifference take hold in our final moments. [2] The dying lie down, stretch their arms toward Jerusalem and receive their last rites: there is a statement of regrets, remembrance of loved ones, the discrete display of emotions, solemn pardon offered to those around them, then prayers and absolution administered by a priest:

> Mentally, this traditional familiarity implies a collective conception of destiny. Familiarity with death is a form of acceptance of the natural order.[3]

Ariès notes that the experience becomes more introspective as the Middle Ages progress into the Renaissance:

Similarities develop between the traditional death bed and the individual's Last Judgement. Whereas previously dying in one's bed is a peace accorded as an essentially collective right [...] On the contrary, judgement [...] enters regarding the indeterminacy of one's fate and its dependence on the weight of the intercessors' souls and prayers. [...] After this point (around the 15th century), each person begins reliving their entire life at the moment of death in a single vision.[4]

The 15th and 16th centuries saw a rupture in a certain primordial familiarity that had normalized human interactions around death, leaving internalization and isolation in its wake. The practice of personalizing burial sites with epitaphs, which sprung up in the 16th and 17th centuries, marked a completely new attachment to the person once deceased, but did nothing to ease their passing.

The 18th and 19th centuries, on the other hand, saw a rise in exaltation and dramatization, a certain preoccupation with the death of others that supplanted one's own, which gave rise to a cult of regret and remembrance that manifested in a proliferation of cemeteries at the turn of the 19th century. Imagery that was once common becomes sparse, and is left to the ancestors, while the deceased themselves are relegated to a periphery and closed off in cemeteries that are visited once a year on anniversaries.

During the second half of the 18th century, the door begins to close even further. Philosophers such as Diderot (1713–1784) and Holbach (1723–1789) deem death unworthy, particularly in light of modern medical triumphs and advances. Death frequently becomes an embarrassing accident about which humans should think as little as possible, because the problem cannot be resolved through science and rational (here again, the experimental method knows failure).

The West in Our Time: Death is Prohibited

We live in an era that systematically does away with anything that can remind us of death. Ariès doesn't hesitate to use the word "prohibited": death has become as taboo as sex. The dying are actually robbed of their own death. Mandatory hospitalization strips them from their homes and beds to die in a sterilized bed and gown, in an atmosphere that permits only a certain degree of emotion, without any attempt to prepare the individual for what might await.

Death is denied at literally every turn. The richest and most "evolved" countries see only an expanding periphery, where cemeteries and crematoriums are run efficiently and hygienically, and the mourning process follows the example.

It's not obvious where to pinpoint the origins of this denial, which now seems possible to characterize as neurosis or psychosis. With the collapse of religions and most spiritual beliefs rendered "unconventional", the remaining post-mortem alternatives find themselves ill-equipped in the face of the only pragmatic option: an uncertain void.

Surveys taken between 1969 and 1977 reveal a sample population's statistical response to the subject of death:[5]

- 46 per cent resentment mixed with fear and anxiety
- 10 per cent with revolt and anger
- 43 per cent with calm

(Notably, 56 per cent have a rather negative outlook.)

- 77 per cent hope for a sudden death
- 53 per cent would prefer not knowing it was coming

As far as the afterlife is concerned:

- 69 per cent believe in God

- 44 per cent believe in an afterlife

Or more specifically:
- 29 per cent of people believe in God and an afterlife, and practise religion
- 20 per cent believe neither in God nor in the afterlife
- 51 per cent sit between these two categories (e.g., they believe in God but not an afterlife, or vice versa)

A Displaced Existence

While the times show a trend toward the complete synonymization of existence and biological life, Jean Jaurès offers a perspective that is both very succinct, very general, and much less anxiety-prone: *death is a displaced existence.*

Before closing our brief history of death, and opening another into modern research into near-death experiences, we'll briefly check in with the schema of the conscious mind proposed in chapter four and place it under the same lens that we have so many others, as such a proposition naturally touches on the topic of death.

Biological functions cease, and the soul of the recently deceased completes its passage through a barrier of light, whereby the soul accesses a more integral and fundamental state of existence in the conscious domain. There, the superluminal material construct of their consciousness endures, living in a spacetime outside temporal law that ultimately says, without paradox: they were *already* dead, they're *still* dead, they *always have and always will be* dead.

For those among us still living, we'll proceed to the next chapter and examine research into near-death experiences carried out by American doctors and psychologists such as E Ross, R Moody, K Ring and M Sabom.

6

NEAR-DEATH EXPERIENCES: THE RESEARCH

By the mid-20th century, the circumstances surrounding the death of a human being had become very different. It was from a hospital room, alone, perhaps even without human comfort or warmth, in an environment sterilized from religion and ritual, that they were now meant to take their first steps into the great unknown. These are the conditions in which each one of us is now destined to confront our ultimate fate, which, paradoxically, is a highly significant event.

However, this turn of events has also had the unexpected consequence of positioning hospital nurses, psychologists and doctors on the now only remaining frontier of death, many of whom have been led by their daily experiences to delve deeper into this primordial phenomenon.

Elisabeth Kübler-Ross: Resuscitating Death

The first study to draw the attention of the public and the scientific world was the 1969 publication *On Death and Dying* by the Swiss-American psychiatrist Dr Elisabeth Kübler-Ross, which statistically analyses behavioural trends in the terminally ill.[1]

While working as a psychiatrist at the University of Chicago's Billings Hospital, Ross honed in on a precise method of interrogation to investigate terminally ill patients. With their permission, she conducted interviews in the presence of the patient's doctor and a chaplain, from which she identified statistical behavioural patterns: namely, five behavioural milestones that individuals tend to experience as death approaches:

- **Thanatic shock**: the individual is traumatized by the news of their terminal diagnosis. Attention is diverted to insignificant and futile activities that aid the individual's denial and unwillingness to confront the reality of their own demise.
- **Anger**: the individual is irritated by those others around them whose lives will continue. At this time, there's often a refusal to take medications in order to frustrate doctors.
- **Depression and bargaining**: an alternation between depression and bargaining with death. The individual promises to be a better person and more tolerant if God allows them to continue to live.
- **Acceptance**: the final stage in which the individual overcomes the depression and, having accepted their fate, finds serenity. Throughout the course of this last phase, the patient is often subject to visions and hallucinations that they attribute to the beyond.

Dr Ross also interviewed patients who had been resuscitated after having been declared clinically dead. The results were quite unexpected: some were able to provide unlikely details regarding their own resuscitation, recalling unique elements of conversation between the doctors and nurses present – all while supposedly plunged into a deep state of unconsciousness that would render sensory function impossible.

Such publications, coming from a highly reputable psychiatric doctor, resulted in a virtual scandal in the scientific and medical world. Nevertheless, today Dr Elisabeth Kübler-Ross is considered a true pioneer in a new field of science: she had temerity to resuscitate death after it had been banished from the minds of the living. Others soon followed suit.

Raymond Moody – NDEs: Near-Death Experiences

The first of these was Dr Raymond Moody, a psychiatric doctor with an MD and PhDs in both psychology and philosophy. As an undergraduate in 1965, he met a professor of psychiatry who had been twice declared clinically dead and reported strange experiences in those moments. Later, while teaching a philosophy course on Plato's *Phaedo* at the University of North Carolina, Moody came across a student's account of a grandmother who had briefly died during a surgical procedure. What struck Moody was the complete similarity of the accounts, given by complete strangers several years apart.

Moody then launched a systematic investigation into individuals reporting NDEs (near-death experiences). In total, he collected around 150: from students, patients, people referred to him by colleagues, and others. In 1975 he published *Life After Life*, in which he categorized NDEs into three separate groups:[2]

1 Individuals that were resuscitated after having been medically declared dead or clinically dead
2 Individuals that suffered grave accidents, injury, or illnesses.
3 Unconscious individuals on the verge of death that gave confirmed testimonies regarding the circumstances of these events.

Kenneth Ring: Rigid Mortality

While Dr Moody entered this new field of research rather serendipitously, pursuing similarities that he'd encountered by chance, sociopsychologist Dr Kenneth Ring entered it in May 1977 with the express intent of utilizing more rigidly standardized forms of data collection and quantitative analysis in NDE research. He launched a three-month study working with a research team at hospitals in Maine and Connecticut.

He used his results to:

- confirm Moody's findings;
- present a more detailed version of Moody's tabular analysis via more thorough statistical evaluation;
- refute criticisms that suggested physiopathology, psychopathology, toxicology and neurological conditions (proposed by several authors);
- further refine the statistical characterization of NDE stages.

Michael B. Sabom: A Statistical Study

At about the same time, an independent study conducted by a young cardiologist, Michael B Sabom, who intended to refute Moody's conclusions, ended up doing the opposite. Statistical analyses of his team's clinical research into NDEs and resuscitation in Florida hospitals were in very close agreement with Kenneth Ring's findings and also further supported Ring's rebuttal of alternative hypotheses.

Statistical Characterization of NDE Stages

The independent research on individuals who had been resuscitated after having been declared clinically dead was conducted by King and Sabom, and enabled the refinement of Moody's statistical characterization of NDE stages. First, let's take a look at the general theme, keeping in mind that the survivors of NDEs each report their own personal experiences which need not conform to, nor contain all the common elements.

- The person is dying. They can hear the medical team, the doctors. They realize that they're dead. They start to hear a distinct hissing or buzzing. They're plunged into darkness. They feel like they're in a sort of dark place (often a tunnel). The person feels as if they're leaving their physical body behind and has an out-of-body experience where they observe the medical team trying to resuscitate them. At scenes of accidents, sometimes the person sees the reactions of other spectators present.

- The out-of-body experience is accompanied by great emotional tension, profound surprise and sometimes anxiety. Then the person starts to familiarize themselves with their new state of being. They notice that they can move through solid matter and travel at great distances instantaneously. They lose their concept of time.

- Others like them start to appear. The person recognizes deceased friends and loved ones. There's a sort of welcoming.

- There's then an intense piercing light, and the person experiences a state of total bliss. They feel that this light is a sort of presence, a being of love. The events of the person's life are spread out in a panoramic snapshot, and each is judged for its true worth.

- The person then crashes into a barrier: a point of no return, a sort of final frontier between life and death. Most of the time, the person wants to cross over. They're overwhelmed with happiness and don't want to return.

- Suddenly, the person feels pushed back. They return to life and reintegrate into their physical body. They regain consciousness.

Another commonality is that these experiences tend to mark the individual profoundly, modifying both their personal, interpersonal and social behaviours. There's also great difficulty in expressing what exactly occurred during NDEs, and a sense that doing so is tantamount to explaining the incomprehensible.

The general theme above was pieced together from individual testimonies taken by American doctors, but is rarely experienced in its entirety, and stages can be absent or overlap. Ring and Sabom therefore conducted independent research into the elements that statistically characterized NDEs. Ring's results suggest the following elements and order:

1   incommunicable aspects
2   individual pronounced dead
3   feeling of calm and peace
4   auditory phenomenon
5   dark zone
6   out-of-body experience
7   meeting others
8   luminous phenomenon
9   judgement
10 final frontier
11 return

However, Sabom's research concluded something slightly different:

1   sense of being dead
2   emotional content
3   feeling of being separated from the body
4   out-of-body observation of material objects or facts
5   empty or dark area
6   judgement
7   luminous phenomenon

The results are quite thought-provoking. But before going down that road, we should tie up some more tangible loose ends regarding our interpretation of the results of these scientific studies, namely: *what is death and how is it defined both biologically and clinically?*

Defining Death

The essential difference between biological and clinical death is that the latter may be reversible, while the former is not; however, a distinct criterion worthy of characterizing cases for medical research is not exactly clear.

Dr Vladimir Negovski conducted experiments for several years in an attempt to both define and push these boundaries. He believed that clinical death should imply the cessation of all exterior signs of life (such as cardiac activity, respiration and reflexes) but not quite the death of the totality of the organism; whereby there remains a possibility of restoring function, and the possibility of reversing a state of clinical death. But if the organism is left to follow its natural course, whereupon the tissues cease metabolic activity, the clinical death becomes biological, and the state is irreversible, with no hope of resuscitation.

This sort of reasonable proposition nevertheless attracts some flak, in particular around distinctions between diagnoses of cardiac death and brain death. For example, a strict adherence to Negovski's definition would imply that a brain-dead individual who retains cardiac function is not clinically dead – despite such a diagnosis being generally perceived as rendering further treatment gratuitous. There's also a tendency for modern medicine to employ the term "clinical death" for a wide range of conditions of varying severity, such as cardiac or respiratory arrest, coma and when the individual is non-reactive after loss of consciousness.

Noting the difficulty in exacting a precise definition regarding criteria in NDE research, Sabom chooses the following approach:

> We've chosen to select patients that we can say have been to that last extremity, that's to say in a physical state that results from a total physiological collapse, accidental or not, where they can be reasonably believed to have bordered on an irreversible biological death, in most cases being called back by emergency medical care when applicable. This definition generally includes cardiac arrest, serious trauma, deep comas arising from metabolic or other organic illnesses, and other similar states. [. . .][3]
>
> In the framework of this study we decided to use the term unconscious to designate the specific lapse during which one loses all conscious perception of themselves and their environment. More commonly, it's what we call losing consciousness.[4]

While Moody didn't analyse his results statistically, Ring and Sabom's studies were both quantitative and independent. Sabom observed NDEs in 39 per cent of patients, while Ring noted 43 per cent. Demographic analyses (such as age, sex, race, profession, education and religion) suggest that these factors play no role in the frequency of NDEs nor their content. Certain individuals even reported aspects of their NDE that contradicted their own religious beliefs. Another study also looked for correlations with individuals having prior knowledge of Moody's NDE research and found that there was no augmentation of the frequency of NDEs, and in fact the effect may be negative.

Looking at the big picture, we see the birth of a new field of research in the United States, with its own protocols, methodologies and statistical analyses that attempt to render data collection and analysis as scientific as possible. To our knowledge, this was the first time that fundamental questions pertaining the frontier between life and death had been approached in this way.

7

NEAR-DEATH EXPERIENCES: INDIVIDUAL TESTIMONIES AND OUR SCHEMA

At this point we've seen two statistical derived characterizations of the most common elements of near-death experiences (NDEs): one presented by Moody and later refined by Ring, the other produced independently by Sabom. In this chapter, we'd like to look at elements of NDE testimonies that exemplify the Moody–Ring findings, and in parallel attempt schematic interpretations that postulate how our materialistic schema for the conscious mind (shared in chapter four) might theoretically be of service in explaining these "unconventional" phenomena. Finally, at the end of the chapter, we'll briefly look at an alternative and more general characterization presented by Sabom, which yields interesting results.

Near-Death Experiences: 11 Principal Elements

Moody and Ring's research suggests that there are 11 statistically identifiable principal components of NDEs.

1. Incommunicable Aspects

The first element identified in NDEs is the incommunicable nature of the experience. He cites a young woman trying to understand this element of her own NDE:

> You see, for me it's a problem to try and explain because all the words I would use apply to three dimensions. During my adventure, I kept thinking: "My geometry classes taught me that there are only three dimensions, which is something that I never took for granted. But they're wrong: there are more." Of course, the world in which we live is three-dimensional, but the other world isn't that way at all. That's the reason it's so hard to explain. I'm forced to use three-dimensional words. I try to stick as much as possible to reality, but that's not exactly it. I can't seem to give you an exact picture of it.[1]

It's worth noting here that our language has indeed evolved in the very particular three-dimensional construct that describes our habitual everyday experiences and – given that the nature of NDEs are anything but – a semantic barrier is likely to exist regardless of the interpretation. Whereas incomplete testimony is often associated with fraudulence, here, on the contrary, it doesn't seem foolish to grant the witness a certain leeway on account of the fact that they might be genuinely dumbstruck by their own experiences.

According To Our Schema ...
The reference to higher-dimensional space and the inadequacy of spacetime are naturally conducive to our schema's conception of the conscious domain, especially regarding the unfamiliarity with what we've postulated to be a uniquely spatial spacetime that lacks any temporal properties. The insufficiency of our language also follows

by the same token, as our methods of communication have evolved uniquely in a three-dimensional cognitive construct. Alternative forms of spacetime would then present a natural semantic barrier, as they have been completely absent during the development of language.

2. Individual Pronounced Dead

The second common element in NDEs is that the individual is pronounced dead, which, in the context of an NDE testimony, means that the individual has somehow witnessed or reached this verdict when in theory any registration or interpretation of sensory information would have been impossible.

Ring cites the case of a woman who suffered internal haemorrhaging two weeks after giving birth:

> In the emergency room, I told myself, "That's it. Goodbye." The only feeling I had was that I was slipping away. I heard them say that I was in a state of shock. I heard the nurse announce: "I can't get her pulse, she's no longer breathing, she's gone." Then I heard another nurse say, "Start resuscitation." But it all sounded like faraway echoes.[2]

Another young woman, who suffered a severe car accident and died three times from cardiac arrest, reports witnessing an entire conversation between her doctor and the surgeon while she was in a state of cardiac arrest and observing her surgeons while she was being operated on. Challenged by her doctors two weeks later, she repeated back their conversation word for word and, even more surprisingly, described the different phases of her operation. Moody cited several other similar cases: individuals declared clinically dead who perceived events that they shouldn't have been able to through normal sensory channels.

According To Our Schema...

Death marks a return to the conscious domain: a domain of higher-order information of which we glimpse fragments through a small aperture that usually remains closed. But in death the mind returns to this conscious domain, passing from subluminal to superluminal spacetime, and in this process the superluminal self becomes increasingly operational.

In regards to the capacity to collect and interpret sensory information when the sensory organs are otherwise presumed to no longer function, we can recall our schematization (chapter two) of sensory experience as a cognitive processing of superluminal information in sync with the manifestation of electrical activity in the cerebral cortex. As the roles of the physical sensory organs are relegated to a reception that lies outside the more complete domain of information, there is nothing that prevents an individual from ascertaining sensory information vis-à-vis the superluminal conscious. This is particularly the case if one embraces the perception of sensation as a form of information. However, the transition from subluminal to superluminal information remains difficult, which as we saw above, corroborates the difficulty individuals have in giving precise testimonies regarding their NDE.

3. Peace and Well-Being

Next we have a phase of peace and well-being, which Ring identified in 60 percent of NDE cases, 71 percent of whom explicitly used the terms *peaceful* and *calm*.

Moody gives several examples in this area. One woman who suffered a heart attack reported:

> I started having wonderful sensations. I felt nothing but peace, comfort, well-being, and a great calm. I had the impression that all my troubles stopped, and I said to myself, "That's so nice, so peaceful, I have no pain anywhere [. . .]"[3]

116

Another man states:

> I felt only an agreeable feeling of solitude and peace. [...]
> It was very beautiful. My spirit was at peace.

A soldier reports that after he was shot in Vietnam he felt:

> A sort of great relief. I didn't suffer and I never felt so
> relaxed. I was completely relaxed and everything was
> good.

Ring cites the case of a woman who, in attempting to commit suicide by drowning herself in the ocean during winter, lost consciousness and felt feelings of warmth, sunshine, peace, security and well-being.

The overall theme among these testimonies, then, seems to be: calm, peace, well-being, the disappearance of worry and fear, and feelings of beauty and perfection.

In contrast, the living human body is in a constant state of communication that utilizes the nervous system, hormones, feedback loops and many other biological factors in order to combat its natural progression toward disorder and decay (the second law of thermodynamics rather pits us against disorder). High-level integration of sensory information coming from both internal and external stimuli is processed in the hypothalamus and constitutes a contributing factor to our emotional state and overall personal experience. This can then be compounded by our daily experience, where we may often encounter sources of stress, anxiety, negativity and anger.

According To Our Schema...

At the moment of death, the conscious mind is relieved of its burden of influx coming from the subluminal domain, that's to say from disorder. New sensations coming from the superluminal domain (that of information, order and a higher consciousness) flood in. They can be nothing but positive and agreeable, because they come from a world where order is constantly increasing and where all is harmony. The sensations of anxiety, fear and excitation – linked to the constant disorder of subluminal bodies – completely disappear. We're left only with the sensations that we find agreeable: warmth, beauty (which is, for that matter, the mark of perfect order, because virtually every work of art attempts to instil a form of order in raw material), well-being and pleasure are admitted to the conscious mind of the individual.

4. Auditory Phenomenon

Moody reports that some individuals heard a distinct auditory phenomenon, which could vary quite a bit from person to person: a harsh buzzing, a loud ringing, a sharp humming, bells ringing, and even very beautiful music. Ring states:

> The memory of these auditory impressions is not only rare but also sometimes uncertain. Most often, investigations reveal that the individual cannot remember having heard strange sounds and responds alternatively: "On the contrary, it was very silent."[4]

One survivor reports:

> Everything happened in a profound silence, the deepest silence I ever knew. There was no sound at all.[5]

There seems to be some discrepancy regarding the frequency that auditory phenomena are reported in NDEs: indeed, it appears on the contrary that the vast majority of individuals recall a complete and total silence. At the same time, it's worth noting that intense sounds have also been anomalously reported by individuals placed in soundproof chambers.

According To Our Schema...

We can interpret the few cases of auditory phenomena according to our schema in the following way: when the conscious mind becomes separated from corporal sensory input, a similar anomaly occurs.

5. The Dark Zone

The "dark zone" is generally perceived as peaceful and is sometimes compared to a tunnel, cylinder, narrow valley or cavern, etc. Moody quotes a man suffering from a serious illness:

I find myself in an empty space, in complete blackness. It's difficult to explain, but I sense that I'm sinking into this void, in complete darkness. However, I'm fully conscious. It's as if I've been plunged in a cylinder without air. I feel in a state of limbo: I'm here and there at the same time.[6]

A woman describing her NDE following a traffic accident states:

It was a feeling of absolute peace and I found myself in a tunnel, a tunnel formed of concentric circles. Later on, after this adventure, I saw a TV show called *The Time Tunnel* where people travel back in time through a spiraling tunnel, and that's the closest image I've found.[7]

Ring mentions another account given by a young woman who'd been in a motorcycle accident:

I felt as if I was ... it's as if I was floating. It's as if I was inside and, believe it or not, the colors are ... there are no colors, it's darkness. This darkness is empty. Yes, that's it: space. Simply a void. But a void with an existence. It's like trying to describe the edge of the universe.[8]

This is notably reminiscent of ancient depictions of passages through dark valleys in the underworld, as well as the archetypical fear that synonymizes death with an eternal dark void. Interestingly, this has a sort of cosmological scale of manifestation with which we're all now very familiar: the black hole.

Astrophysics describes a collapse in certain stars that can occur when gravity compresses their radius to the order of about 10km. Spacetime becomes so distorted by the enormous density of the gravitational field that it forms a sort of pocket that folds in on itself and from which nothing can escape, including light.

Equally interesting is that some cosmologists have actually looked at modelling our entire universe on the interior of a black hole. Régis Dutheil studied this using superluminal relativistic theory, which resulted in a universal black hole having a radius of several tens of billions of light-years (outside of which existed superluminal spacetime). And what's even more interesting is that Dutheil, along with J P Vigier, used general relativity to envisage electrons as micro black holes with superluminal interiors.

According To Our Schema ...

In fact, the entire world that surrounds us, from tiny electrons to colossal galaxies, would be a double: there's always an upside down in a set location. What we see every day is subluminal, but as soon as our conscious is relieved of the filters that habitually deform and obscure part of reality, it can perceive the other part of the universe: the superluminal domain.

However, going from one domain to the other requires crossing a barrier: the light barrier. It's there on the surface of electrons, separating the superluminal interior from the subluminal exterior, and it's there on the frontier of our known universe, separating us from the superluminal exterior.

The dark zone is then easily understood. Crossing the dark zone corresponds to the conscious mind crossing from the subluminal domain (the interior of a black hole) to the superluminal domain: crossing the light barrier. At the barrier, the conscious mind enters the luminal domain (or light barrier), whereby it becomes imbued with luminous particles – at which point the exterior can appear as nothing but darkness. Individually, this resembles crossing a dark region of space comparable to the edge of the universe. From this perspective, death now involves more than one collapse, as we're now dealing with black holes.

6. Out-of-Body Experience

Out-of-body experiences can be generally characterized as a form of perception that exists outside normal channels of sensory processing. Moody cites two particularly interesting testimonies. First, a woman who states:

> I felt myself leave my body and slide down between the mattress and the bed railing. It seemed as if I went straight through the railing, actually, right to the floor. Then I gently lifted myself up in the air... I continued to lift myself up through the ceiling, looking down. It felt like I was a piece of paper that could fly simply by blowing on it.[9]

This next out-of-body experience occurred outside the hospital, and is from the testimony of a young man who was involved in a car accident:

> I found myself floating around 1.5 yards above the ground, around 5 yards away from the car, and I heard

the echo of the collision grow further away and then fade out. I saw people running up and gathering around the car... I also saw my own body among the wreckage, with people around trying to pull it out.

Many individuals report being dumbfounded, both on account of a sadness from having to leave their bodies and a confusion regarding the state in which they find themselves. Others experience serenity or an emotional detachment. The overall tendency toward anomalous physical properties is quite striking.

Onlookers came running from each direction to the accident. I watched them from the middle of a very narrow sidewalk. However, when they got nearer, they didn't seem to notice me. They kept walking, looking straight ahead. When they got close, I tried to get out of the way, but they walked right through me.[10]

Another survivor describes:

My being had a certain density, well, almost. Not a physical density, I'd say more so like waves or something like that. I don't know, nothing like real material, it was like an electric charge, if you like. But it was still something. It was small, vaguely spherical, but without a precise contour, barely a cloud [...]. All of it was very light, very light. There was no tension on my (physical) body. It was a feeling of total separation. My body was lighter than air.[11]

This description of energy is not unique:

I would best describe it as an energy centre... It was like I was there – an energy perhaps, or if you like, a small concentration of energy.

One survivor reported a hyper-developed intellect:

> My spirit became wonderfully clear. I noticed everything and resolved all problems like never before and without having to think about things twice.[12]

Another woman mentions an ability to travel at great distance:

> When I wanted to see someone who was far away, it was if there was some part of me, a kind of honing beacon, that launched me toward this person. And it seemed that wherever anything was happening in the world, it was easy to go and see it.[13]

And yet another survivor speaks of a distinct form of extra-sensory perception:

> I saw people around me and understood what they were saying. I didn't hear them in an auditory way, as I hear you. It was more as if I knew what they were thinking, exactly what they were thinking, but only as an idea, not their vocabulary. I captured their thoughts a second before they opened their mouths to talk.[14]

Finally, a number of testimonies collected by Ring mention a total alteration of the perception of time during the out-of-body experience:

> I no longer had any concept of time. Time no longer meant anything.[15]

A second testimony adds:

> I had no idea how much time went by. Sometimes when I look back, it seems like it was an eternity.[16]

A third:

> What was interesting there, was that everything passed outside time and space. That was necessary, because the context is simply... one cannot classify things temporally...[17]

A fourth:

> I found myself in a certain space, for a certain time, while, I dare say, the very concept of space and time were abolished.[18]

According To Our Schema...

The similarities in these testimonies are self-evident. First, the conscious of the individual has access to all the usual sensory interpretations – which could even be hyper-developed – and finds themselves localized outside the body (for example, the individual who stated that he was 1.5 metres above the ground and five metres from the side of his car). Second, there is an altered perception of time and space, while intellectual faculties are rendered extraordinarily acute. In the majority of cases, the individual feels like they're a whole new being, a kind of energy, something much less dense than ordinary matter. The conscious mind seems capable instantaneous travel, anywhere, and the perception of time is also altered.

These testimonies give us a glimpse into a conscious mind that's penetrated another form of spacetime.

According to our theory, it's during this phase that the individual's conscious is imbued with light particles (photons) after having traversed the light barrier and where it would necessarily be formed of *solitons*: photons or neutrinos that exist on the light cone. During the sixth phase, the luminal and superluminal parts of the conscious mind detach from the physical body, which gives individuals the feeling of a pure state of consciousness with an exceptional acuity (because the superluminal domain is one of pure and total information). The solitons that impregnate the conscious mind belong to their own spacetime, which explains the altered perception of time and space (remember that time doesn't flow in the superluminal universe). Solitons are formed from a network of stationary waves, which explains the reported sensation of energy, as well as localization at various heights.

The conscious mind of the individual then passes through the light barrier and prepares itself to enter the superluminal universe, where it begins to experience the first effects. The actual passage occurs during the seventh and the eighth phases, which often overlap.

7. Meeting Others

During this stage, individuals report being reunited with lost friends and loved ones. Moody cites a man who saw one friend in particular:

> When I left my body, I quickly felt that Bob was right there beside me. I saw him mentally and sensed he was there, but it was a curious feeling: I didn't see him physically, even though I could distinguish the rest of my physical surroundings very clearly, their features, everything. I'm not sure if I'm explaining myself well. He was there, but he didn't have his usual body. His body was transparent. It seemed like he had all his limbs, arms and legs, but I couldn't say that I saw them physically...[19]

The man goes on to say that he anxiously asked Bob if he was dead or if he needed to leave, but received no response. Some report that this line of questioning is answered in the affirmative – *that they are dying and would soon feel perfect* – and others recall meeting people with whom they were not familiar.

8. Luminous Phenomenon

This phase involves an intense light that's sometimes associated with a warm and reassuring presence, which is often experienced in lieu of meeting others.

Moody quotes a young woman:

> I continued to float through the beams as if they didn't exist, and from there toward this light of pure crystal: a beaming white light, very beautiful, very bright, irradiating. But it didn't hurt my eyes. It was unlike anything I'd ever seen. I couldn't say that I saw a person in this light, but it seemed certain that it possessed an identity, that's undeniable. Imagine a light made of total comprehension and perfect love. A thought came to mind: "Do you love me?" [...] And during this time I felt completely surrounded by compassion and overwhelmed with love.[20]

One man attempts a more precise physical characterization:

> In the beginning, it seemed a bit pale, but all of a sudden there was this intense beam. The brightness was remarkable. Not lightening, but a blinding light, that's all. And it gave off a warmth, I felt warm. It was shining white, perhaps a little yellow, but above all white. It was incredibly bright. I can't manage to describe it. It lit up everything around me, but didn't prevent me from seeing everything else: the operating room, the doctor, the nurses, everything.[21]

Ring interviewed another woman that describes the involvement of more colour:

> The light is very, very bright, as if the Sun itself was in the room. You could say that the colours were all bright.

You know, as if all the light was amplified... (the colours were) perfect with respect to their inherent color. [22]

Ring also showed statistically that the perception of light is one of the most advanced stages of NDEs. Of the 60 per cent of testimonies that report experiencing feelings of peace and beauty in phase three:

- 37 per cent report out-of-body experiences
- 23 per cent enter the dark zone
- 16 per cent witness a kind of luminous phenomenon

He also found that just 10 per cent reported having penetrated directly into this light, which he concluded was one of the deepest phases of the experience, and cites several cases that mention beautiful landscapes.

One woman interviewed by Ring states:

I found myself in a field, a vast and deserted field with tall, golden plants. It was very calm and bright. [...]. The plants were so extraordinarily beautiful that I'll never forget them. [23]

Another man reports:

I saw flowers there that no one on Earth has seen before. I don't think that there exists a single color on Earth that wasn't represented among the colors that I saw. [24]

According To Our Schema...

To interpret these different aspects of the testimonies, it's necessary to recall this essential fact: superluminal spacetime represents a universe of information, meaning and order (thus harmony) that constantly increases. It contains both the events that an individual can come to know during the course of their subluminal terrestrial life, and an incalculable amount of information that the deceased enters into – that of light. In addition, there is no flow of time in the superluminal domain: all events occur simultaneously.

We've already seen our schema show that our everyday reality is an assembly of holograms produced by the conscious mind. While the superluminal or conscious domain consists of information in a pure state, it's also reasonable to assume that a deceased individual arriving in this universe would continue to translate this information into holograms for a certain period of time.

All of this explains why the deceased perceives the presence of their close ones and why they seem to have translucent forms. As the filtration processing enacted by the cerebral cortex and sensory organs no longer exist, the individual's experience of the conscious mind's holographic projections is altered: in a sense it is purified, more direct, and closer to the information's raw state.

The eighth phase appears to individuals as an extremely bright light with wondrous colours. As superluminal matter is necessarily much less dense than its subluminal counterparts, it has less of a capacity to absorb radiation (e.g. light, photons), which explains the overall resulting brilliance.

In addition, these light particles should probably carry information and meaning, which results in the peacefulness and significance associated with this light. The supernatural purity of the colours would rather be due to the fact that the sensations experienced by the individual are no longer attenuated by the cortex and arrive in a raw state.

The wondrous landscapes, seen in some testimonies, are nothing more than holograms constructed by the conscious mind to translate incoming information. These holograms[25] are in some ways idealized replicas of subluminal counterparts that can both manifest and annihilate instantaneously, which is reminiscent of one of the essential teachings of *The Tibetan Book of the Dead*: the mind of the deceased creates their environment after they die.

Of course, our Aristotelian logic still encounters certain problems with a false reality. Subliminal reasoning would say that these objects and landscapes aren't real, because they can appear or disappear at will. On the contrary, we'd say that a subluminal object or landscape is real precisely because they can't. But we forget that these subluminal objects are holograms, and therefore not real, and that we ourselves (our bodies) are nothing more than holograms separate from our total consciousness, which is also not real. It's therefore rather the unreal within the unreal that gives the impression of reality, like two negatives making a positive. The only real thing is the superluminal conscious mind that can create the universe at will.

9. A Panoramic Vision

Judgement, particularly in the form of a panoramic of one's life, is particularly reminiscent of religious archetypes.

Moody cites a very detailed account given by a young woman who claims that a luminous being (the eighth element) asked her to judge the constituent events of her own life. She saw herself as a child breaking a favourite toy, then, later, as an adolescent in high school.

> I was there, I really saw all these scenes. I went through them and it all happened so fast; however, there was enough time to not miss anything, and yet on the whole it didn't last long. At least, it didn't seem that way. The light came first, then looking into the past, then the light again. I'd guess it took somewhere between 30 seconds and 50 minutes, but I couldn't tell you more beyond that.[26]

However, both Moody and Ring found that this panorama is not necessarily associated with the being of light. One of Ring's interviewees states:

> It wasn't exactly images, I'd say more like forms of thought. I don't know how to explain it, but everything was there, everything was there at the same time, I mean, not a succession of pictures shining one after the other, but rather a mental view of the whole set at the same time.[27]

Another survivor describes:

> The return took the form of mental images, but the images were much more alive than in normal time. I only saw the important moments, but everything happened so fast, as if I looked through the entire book of my life in just a few seconds. It happened like a high-speed film, while still being able to see and understand everything.[28]

A young veteran of the Vietnam War states:

> The best comparison that comes to mind would be a slide show that someone was showing at high speed.[29]

One young woman interviewed by Ring states that she saw her future as well:

> It's as if I saw my husband at the same time as an image of us five years later. I saw myself with our children. You could say that I saw and that I knew the children that I would have.[30]

She saw a future as the mother of two boys, and indeed that ended up coming to fruition.

According To Our Schema...

During this ninth phase, a dialogue occurs between the subject and the entity that takes the form of an instantaneous panoramic of one's life (or parts of it).

This is actually a dialogue between the subliminal self and the total superluminal consciousness, which is infinitely richer in information and knowledge. We already noted that the superluminal conscience can instantly access all information regarding the individual's past, present and future. Therefore this higher self has no difficulty presenting, in a domain outside of time, a reel of events marking the past, present and future milestones of the individual in the subliminal universe. The projections of these events comprise holograms,[31] exactly like the magnificent landscapes that appeared during phase eight.

The superluminal conscious benefits from a more direct perception than its subliminal counterpart, because it's not hindered by filters (such as the cerebral cortex and organs). There's then nothing surprising in individuals feeling a special acuity and relief in the depiction of life events. One of the testimonies estimates the duration of this panorama to have lasted between 30 seconds and 50 minutes: as superluminal speeds imply an instantaneous unravelling of events, and a time that doesn't flow, these seemingly outrageous events can be perfectly explained.

It's essential to note that this judgement isn't accompanied by any sense of guilt. There is nothing that brings into question the concepts of good and bad in religion, nor any serious moral revision. On the contrary, the higher self alternatively highlights the importance of knowledge and meaning.

Events judged good or bad in the subliminal world take on a rather transcendental meaning in the spacetime of the conscious mind, which probably has little or no connection with their subliminal counterparts. We can even wonder if the importance of love, highlighted by the higher self, isn't the result of a complex processing of information concerning several selves in the superluminal universe. In a way, everything happens as if a part (the very thin part of the conscious mind that remains accessible in the subluminal word) talks with the whole (the total superluminal conscious).

10 and 11. The Final Frontier and Return

The last element to characterize NDEs in statistical terms is a final frontier, which individuals find themselves unable to cross, and then a return.

Moody interviewed a woman who describes being among beautiful scenery, a meadow of intense greenery and light, when she suddenly crashed into a fence which, despite her efforts, she was unable to cross. Another man stated that they were on a small boat, in the middle of crossing a very beautiful river, where his deceased parents were waiting on the other side to welcome him. However, his vessel did a sudden U-turn and he never reached the shore. A third individual reported a grey fog that he was meant to cross, while a fourth saw a closed door through which he perceived an intense light.

However, it's also very common for NSE accounts to make no mention of any such final frontier. The last element of NDEs is then a sort of return, which can be quite sudden. One man states:

> I was up there near the ceiling, I watched them performing medical treatments on me. When they put the electrodes on my chest, and my body jumped, I saw myself fall like dead weight. An instant later, I was in my body.[32]

A woman explains: "It seemed like I was called back, almost magnetically."[33]

Individuals sometimes report that they found themselves quite happy where they were and didn't want to return:

> I wanted to stay where I was. Then, all of a sudden, I heard my daughter and my children and I understood that I should, I should return.[34]

Ring, Moody and Sabom all cite testimonies in which individuals insist on the reality of their experience: they avidly reject the idea that it could have been dreamed or hallucinated, and emphasize the profound repercussions on their conception of life, death and on their personality; many no longer fear death, feel stronger, more optimistic and calmer and have a more positive outlook on life.

According To Our Schema...

When analysing this portion of the testimonies, we notice that they very often contain an element of symbolism related to Jungian archetypes: the idea of encountering an impasse, which one crosses or does not cross.

The partial subluminal conscious of the deceased searches, throughout this experience, to penetrate the superluminal universe and to join with the total consciousness. Likely, this fusion first requires a certain exchange of information, which fundamentally characterizes the superluminal domain.

If the quantity of information accumulated by the partial consciousness isn't sufficient to stay in the superluminal domain, it's sent out and must return back through the light barrier to come back to life in the subluminal domain: just like feeding insufficient data into a computer and drawing an error response.

On the contrary, if partial consciousness possesses sufficient information, then it is accepted, the individual remains in the superluminal domain, and they die.

In both cases, we believe that there exist unknown physical laws that depend on the exchange of entropy/negentropy, or equivalently disorder/order, between the two types of space.

An Alternative, More General Characterization of NDE Components

It's important to remember that the above characterization of NDE components, first proposed by Moody and later statistically refined by Ring, is far from rigid: one or more elements may be missing, and their order can be inverted. While quite similar on the whole, Sabom's results actually suggest a slightly different order:

1   sense of being dead
2   emotional contentment
3   feeling of being separated from the body
4   out-of-body observation of material objects or facts
5   empty or dark area
6   judgement
7   luminous phenomenon
8   accessing a transcendental world
9   meeting others
10 return

Sabom also grouped NDE components into *autoscopic* and *transcendental* phases.

The Autoscopic Phases: 1 to 4

Sabom defined autoscopic phases as those involving either:

- a feeling of bliss associated with the experience of dying, or
- an out-of-body experience involving the observation of material facts (such as doctors or resuscitation)

These generally correspond to the first four phases identified in his results.

The Transcendental Phases: 5 to 10

Sabom then characterized transcendental phases as those involving:

- the dark zone
- the luminous phenomenon
- meeting others
- judgement

These generally correspond to phases five to ten as identified in his results.

This new characterization offers the following interesting statistical perspective:[35]

- 33 per cent experience exclusively autoscopic phases.
- 48 per cent experience exclusively transcendental phases.
- 19 per cent experienced both types of phases, and in a more or less continuous order.

According To Our Schema...

According to our theory, the autoscopic phase would arise from the fact that the individual's conscious mind is still plunged in the subliminal world: it's simply severed from its connection with the physical body, thus freed from the filters in the cerebral cortex and capable of perceiving incoming information with a much greater acuity.

In the course of the transcendental phase, the conscious mind crosses the light barrier (that is, the dark zone) and penetrates progressively deeper into the superluminal universe. It's normal that some individuals only experience the first part and do not have enough time to cross the light barrier. Others, on the contrary, experience the entire process.

It's also perfectly conceivable that those in a state of partial conscious can suddenly cross the light barrier, without warning, which explains why certain individuals only remember the second part of the experience.

8

OUR SCHEMA AND THE NEED FOR A NEW PHYSICS TO UNDERSTAND DEATH

Our schema interprets death as a purely physical phenomenon, but the relevant physics is yet to be understood. We've already seen modern physics uncover strange new horizons regarding the nature of reality, with particles in the quantum domain behaving very differently from their more familiar macroscopic counterparts. Particles are no longer objects whose colour, shape, dimensions and location are determinate. They also possess a wave–particle duality whose complexity is compounded by the role played by the conscious observer in quantum experimentation. These contradictions, which seem to lack any practical resolution, demand that we expand the arena of physics to encompass the new study of superluminal particles.

Our schema presents a universe that is only half-visible, half-known. Much of it lies beyond the light barrier, in the superluminal domain, which is currently out of reach. Its theoretical study began 20 years ago; however, experimental investigation has proven extremely difficult, to the extent that recent Belgian experiments foresee persistent obstacles. It's only by turning to a new physics that we can attempt to understand death.

In the preceding chapter, we've unravelled the details of many near-death experiences (NDEs) and used our schema to interpret a physical basis of these phenomena; the ease with which we were able to support various observations speaks to the credibility of the model we're proposing. To recap, our schema proposes the existence of two complementary domains: the *subluminal*, which we live and experience day to day, and the *superluminal*, which is only perceived at the moment of death.

This superluminal domain is also that of the conscious mind. We are the image of this superluminal construct: our physical body with all its organs belongs to the subluminal domain, while our conscious belongs to the superluminal domain (which recalls the ancient conception of reality as a microcosm having a macroscopic counterpart). Our everyday conscious experience is, then, rather like living within the interior of a house, outside of which lies an entire new domain of the conscious mind.

This in turn invites the question: how would it be possible for a proverbial speck of dust, which finds itself randomly located in the interior of such an unfathomably large house, ever to begin to comprehend the nature of what lies outside? For human beings, how could we do this while our observations are limited to the physical world that our bodies inhabit?

Some astrophysics even pattern our universe as a giant black hole that neither light nor information can penetrate. The particles that make up our physical body, including the brain and organs, can also be considered micro black holes. The superluminal domain, and therefore the total conscious mind, would be situated outside this black hole, and the barrier between this black hole and the rest of the universe would be marked by the light barrier.

Over 2,000 years ago, Plato provided a captivating illustration of our universe in his allegory of the cave in his

Republic, Book VI, which possesses a dualistic nature that mirrors Régis Dutheil and Vigier's subluminal–superluminal formulation of the electron. The frontier forms at the surface, a sort of mini light barrier, and calculations theoretically show a continuum between one side of the interface to the other. As these phenomena occur at the quantum level, where we find numerous other curious idiosyncrasies posed by quantum mechanics, they remain invisible in our daily lives.

Tripartite Construction

In other words, our schema proposes that the entire universe, from the smallest of particles to the most colossal galaxies, has a tripartite construction reminiscent of so many others since the dawn of antiquity: trifectas of deities, social castes (in particular of the Indo-European variety described by the philologist George Dumézil), feudal society partitions, nobility and the Church (both with respect to its societal roles and in the Holy Trinity).

Specifically, the domains of our schema are distinctly characterized by the speeds of their constituent particles: the *subluminal domain* (in which particles travel at less than the speed of light), the *luminal domain* (where particles travel at exactly the speed of light), and the *superluminal* or *conscious* domain (in which particles travel faster than the speed of light).

Light therefore acts as a sort of reference point, which similarly reflects its prominent role in so many major religions: deified as the Sun in ancient antiquity, spiritualized in Christianity, etc. The importance that individuals accord light in NDEs is similarly indicative, in particular given that these experiences can be interpreted as one of our only means of contact with the superluminal domain.

Making Contact with the Superluminal Domain

According to our theory, the autoscopic phases of NDEs establish a link between the conscious mind in the subluminal domain, while at the same time the normal interactions between the conscious mind and the cerebral cortex cease to function one after the other. The result is a very curious out-of-body experience that's associated with a sense of well-being, which is to be expected, as through deliverance from the body the conscious mind ceases to experience any painful aspects of physical life.

In the transcendental phases that follow, the conscious mind then crosses the light barrier (the dark zone) and breaches the superluminal domain where it's assailed by a new incommunicable form of sensory experience.

At this point one can ask: if it's so easy to understand NDEs, can we know what happens after?

The Domain of the Dead

NDEs, by definition, only represent the beginning part of the voyage: a U-turn is necessary, otherwise we arrive at a rather unfortunate outcome! However, we can still probe into the superluminal conscious domain through hypotheses, scientific modelling, thought experiments, extrapolations and analyses.

First off, it's important to keep in mind that the conscious domain constitutes a profound and fundamental reality, of which our universe is nothing more than a hologram, diminished and altered like a signal passing through a radio antenna.

Next, it's necessary to recall that the conscious domain is superluminal and involves speeds so significantly greater than those we can imagine, that the flow of time loses its meaning and events effectively occur instantaneously. The conscious mind, in this universe, can instantaneously access information on the totality of the events that comprise a human life, and probably much more (many purely superluminal events are probably not projected into the subluminal domain, for reasons that we don't understand).

In addition, it's possible that all superluminal events, which exist in the form of pure information, have an inherent meaning in the conscious domain that lacks any subluminal parallel. In other words, the very essence and the meaning of reality that the superluminal conscious mind experiences is in all likelihood totally different than its subluminal projection.

According to our theory, this projected image or hologram, which is the point of reference of our subluminal universe (because we ourselves are holograms), is insignificant in the superluminal domain, which occupies itself with *something else*. It's therefore necessary to shed, with great difficulty, every anthropomorphism in order to glimpse this other domain, which would be, in fact, *reality*.

Quantum mechanics offers us a very vague idea of what this domain entails: a universe where objects can no longer be localized and where the conscious mind participates in physical reality. The American physicist George Gamow, celebrated for his brilliant work on quantum tunnelling, wrote a very funny book in the 1950s entitled *Mr Tompkins in Wonderland*, which describes the results of quantum phenomena infringing on our everyday macroscopic world.

One evening, a bank employee called Mr Tompkins decides to attend a conference on quantum mechanics. That night, Tompkins has all sorts of nightmares about quantum phenomena popping up in his everyday life. He's able to pull into the garage without having to open the door, as non-zero probabilities allow both his car and himself to quantum tunnel through the barrier. He goes on a safari in India, hunting a tiger on elephant back, but when he takes aim at the tiger it turns into thousands of *potential* tigers: a playful illustration of the strange fact that in quantum mechanics one cannot predict the results of a measurement (here made with the scope or bullet), prior to its having been made.

Gamow's work gives us a small glimpse into our own brave new world, and that's without touching on the how superluminal materialistic properties would play out with cars and tigers.

The Forging of Inconceivable Concepts

Indeed, the conscious domain's immersion in superluminal spacetime renders our problem much more complex than that of Gamow. To paraphrase Hegel, we need to undertake: *The forging of inconceivable concepts.*

Some key words pertaining to the superluminal domain are: *altered spacetime*, *purely spatial spacetime*, *instantaneity*, *increasing order*, *meaning*, *information* and *synchronicity*. We can attempt to imagine an entity that can access two forms of time:

- **An experienced time** (or proper time) that doesn't flow and allows someone to be at any point in their existence instantaneously (as there is no past, present, or future).
- **A cinematic time** that undoubtedly has a value that we can't comprehend, but is nevertheless more than a simple parameter. It's with respect to cinematic time that superluminal entities exist and evolve, even though it doesn't flow. It's a form of evolution dominated by a constant augmentation of order and information, which makes this domain comparable to a sort of river flowing back to its source: order appears increasingly structured and beautiful, with information and meaning, and forming a network of synchronicity whose complexity and magnificence are beyond measure. Our partial subluminal consciousness is fundamentally causal, and it's only on rare occasions that it can glimpse these remarkable laws of synchronicity.

Such a river isn't fixed, like the immutable beauty discussed by Rabindranath Tagore, but is rather an evolution in instantaneity and immutability. And here we see that the time has come to abandon Aristotelian logic and to embrace a wider non-Aristotelian approach that's already been foreshadowed by quantum mechanics.

It's clear that such a universe is fundamental in both meaning and information. Many survivors of NDEs insist on the importance of knowledge. Some had the clear impression that they continued to learn beyond death, or that certain information would be necessary to enter into the beyond.

Nobert Wiener, one of the pioneers in informatics, concluded that it was actually impossible to give the abstract concept of "information" a concise definition. Information is, rather, a fundamental quantity, like energy, and similarly eludes precise definition. We can therefore imagine superluminal structures, which have long ago evolved beyond their holographic representations, as informatical structures on an evolutionary path toward an ever-increasing order that probably coincides with the inherent concept of beauty.

There's then nothing surprising in the emphasis that NDEs place on the importance of love and beauty. Here the term "love" should be interpreted in a very general sense, free from familiar connotations that weaken its intent. In the superluminal spacetime of the conscious mind, every subluminal thought and act of love might correspond to the appearance of new informatical associations, an actual seed of synchronicity that develops a particularly harmonious order among its structures. Beauty is naturally the aesthetic base of that order. For example, a sculptor takes raw material and imposes form and organization, a painter does the same amidst an anarchy of colours, and a musician produces orchestrated harmonies from what would otherwise be scattered noise. This is precisely what would happen spontaneously in the great river of information in the superluminal domain.

We're now approaching a description of the universe that can be likened to Leibnitz's 17th-century schema (see chapter one): each portion of the universe, animate or inanimate, has access to a part of the conscious domain, because each particle (electron, for example) possesses a superluminal quality representative of this domain of information.

One of the essential differences that separates the subluminal and superluminal domains is this: while information in the subluminal domain follows causal sequences (information is linked following causality, which in turn depends on the flow of time), the superluminal follows synchronicity.

But to go beyond this, to delve deeper through the light barrier, involves semantic and conceptual obstacles that are very difficult to overcome. It's likely that at certain times the answers we seek are reflected in certain mysticisms and religions, whose subject matter so often eludes exact expression and translation. Perhaps there are even abstract 20th-century paintings and compositions that offer glimpses into this beyond (it's worth noting that abstract painting seems to have developed during the same epoch as relativity and quantum mechanics).

For those who still prefer a visual representation, one of the best examples is probably Stanley Kubrick's movie *2001: A Space Odyssey*, which was based on the novel by Arthur C Clarke. The plot: an extraordinarily powerful source of radiation is located on one of Jupiter's satellites and appears to correlate with the appearance of a mysterious black monolith on the moon. A spaceship is charged with a secret mission, and only a single member of the team survives to make it to their destination. He meets his death as well, and in the end is reborn as an invisible entity. The corresponding chapter, entitled "Through the Star Gate", illustrates Clarke's main idea: there's another universe, with another spacetime, which represents a sort of afterlife.

While passing through a great portal, the astronaut notices that time stops ticking, then crosses a great black pit to arrive in a new and unfamiliar world:

> There's no atmosphere, as each detail appears in total clarity even unto the incredibly far off horizon. The dimensions of this world must be fantastic. [...] Every expanse that Bowman could see was tessellated into obviously artificial patterns that must have been miles on a side. It was like the jigsaw puzzle of a giant that played with planets; and at the centres of many of those squares and triangles and polygons were gaping black shafts – twins of the chasm from which he had just emerged.[1]

The description of the sky is even stranger and confirms the first astronaut's impression: an alteration of spacetime:

> Yet: the sky above was stranger – and, in its way, more disturbing – even than the improbable land beneath. For there were no stars; neither was there the blackness of space. [...] The sky was not, as he had thought at first glance, completely empty. Dotted overhead [...] were myriads of tiny black specs. [...] These black holes in the white sky were stars; he might have been looking at a photographic negative of the Milky Way. [...] It seemed that Space had been turned inside out.

We find here that familiar idea: the inversion of spacetime.

Conditions of Existence in the Domain of Death

To draw this chapter to a close, we'd like to return to consider what sort of conditions would govern one's existence in the realm of death (i.e., in the conscious, superluminal domain).

During one of the last phases of NDEs, some survivors report having seen marvellous landscapes (fields, gardens, rivers) that correspond to a profound notion of beauty. It's amidst these landscapes that individuals often report meeting family members or entities that guide them toward a new world and encourage them to undergo the judgement of their former terrestrial life. These rather stereotypical representations are, according to our schema, holograms created by the very conscious mind of the individual. We can even go further and hypothesize that these representations are actually Jungian archetypes, in the sense that they're pre-existing models that lie at the forefront of superluminal consciousness.

Western religions (such as Islam, Judaism and Christianity, as well as many other ancient religions) tend to portray a homologous realm of representations (for example, a verdant paradise, hell and places of final judgement). For these religions, such representations belong to the domain of God. Eastern religions, on the other hand, and Hinduism in particular, tend to embrace a multifaceted voyage of the spirit, whereby associated landscapes (paradise or inferno) depend entirely on the state of the deceased's spirit, which creates these experiences through thought. Our physical model is then in complete agreement with these latter conclusions.

This phase would be luminal, from our point of view; that's to say that it occurs at the moment when the conscious mind would be imbued with luminal particles after having traversed the light barrier. All earthly motivations (hunger, thirst, fatigue, pain) are abolished. The deceased has the sensation that they are becoming a pure spirit. But, being still attached to the subluminal domain by a number of links, there is the need to create an environment that conforms to terrestrial life; hence the landscapes, fields, rivers, etc. However, this is simply a phase, a crossing prior to integration into the great superluminal river of information and meaning that is the conscious domain: the true beyond.

One last problem remains: *are all individuals who reach the luminal phase permitted to proceed into the final stages and cross into this superluminal river?*

Entering the Domain of Death: Knowledge

Everything would appear to indicate that this is not put to a vote, not based on religious notions, but rather an evaluation of our physical schema. In fact, it seems that it's the quality and density of information and meaning that determines the fate of an individual's consciousness. If we apply a weight to this information and meaning, and that weight is insufficient upon arrival at the luminal stage, integration into the great superluminal river is not possible.

What happens in this case? The conscious remains in a luminal state, where they retain the capacity to transform information and meaning into holograms that correspond to our subluminal universe. These are then retained as memories by the individual, which can constitute very pleasant landscapes for some (who were honest and good during life, regardless of their not having attained a sufficient quality and density of information to pass through to the final stages), and less pleasant for others.

It's clear that in our schema, moral criteria aren't the only deciding factors, but that knowledge – understanding sought during the span of an entire life – is essential. This agrees with the emphasis that many NDEs place on the importance of understanding in accessing transcendental stages.

True understanding wouldn't consist of erudition in its own right, nor hyper-analytical specialization (unless used as a stepping stone), but would be more of an esoteric understanding that can be interpreted as a rather mystical union of scientific comprehension and the search for the metaphorical Holy Grail. Meditation is clearly essential, because it allows us to empty our minds; to imagine new forms of organization, order, meanings and synchronistic relationships; and to foster abstraction; and regularly increases the amount of information and organization in the conscious mind. Meditation is then rather like astronaut training to experience new conditions of weightlessness. To quote Jung on reflection:

The first part of a person's life should be dedicated to the Earth, the second to the sky.

9

PARANORMAL PHENOMENA

Where there's research into death, paranormal phenomena are not far behind: apparitions, ghosts, possessions and messages from the beyond. The status quo in science is to treat this with a bit of contempt. Let's try to dive in and see if we can prevent ourselves from being completely submersed in superstition.

Apparitions

We can say that there are two very general types of apparitions: those of the living and those of the dead. The exact nature of their manifest reality can be quite imprecise. Sometimes reports indicate phenomena that appear to possess all expected material properties, but which can pass through physical matter. Some phenomena are observed by a single individual, even though others might be present, while at other times several witnesses will corroborate the same events.

There are legions of ghost stories, but very few come close to meriting serious attempts at authentication and investigation. Indeed, the few that withstand the test of time do so because they continue to disturb the most determined minds pitted against them.

The Concert of Corpses

This well-known account has attracted a lot of attention over the years due to the character of the principal witness, the thorough investigation performed by the commissioner in the town of Vaugirard, and the fact that the case remained open for a long time in a prefecture of the Parisian police.

It's June 1925. A 24-year-old man named Jean Romier, a medical student who's generally perceived as having a good reputation, meets an older gentleman in the Jardin du Luxembourg who introduces himself as Mr Berruyer. They have a friendly conversation about music and Mozart, and Berruyer invites Romier to attend a concert that he and his family regularly give on Fridays in their apartment on Rue de Vaugirard.

That Friday, Romier arrives at the Berruyer residence and spends a pleasant evening listening to performances of Mozart, being fed hors d'oeuvres by Mrs Berruyer, and chatting with a young seminarian and the two Berruyer boys, one of whom is attending law school, while the other is intending to go into the navy. Romier leaves around midnight, but he doesn't make it far when he realizes he has forgotten his lighter. He turns back and rings the doorbell.

No one replies. He keeps ringing the bell. He ends up waking a neighbour, who informs the concierge. A police officer arrives and arrests Romier. Both the neighbour and concierge confirmed that the apartment had been empty for years, following Mr Berruyer's death. It took several hours for Jean Romier to convince both of them, along with the police officer, that he wasn't lying.

The next morning, the current owner of the apartment, Mr Berruyer's great grandson (who would have performed music at the concert attended by Romier the previous night), agreed to unlock the apartment door. They found an abandoned apartment filled with spider webs. However,

Romier surprised everyone with his perfect knowledge of the layout, as well the accuracy of his knowledge of Berruyer family members: the young law student matched the description of the current owner's grandfather (a lawyer), the prospective naval student matched the description of his great uncle (an admiral), and the seminarian would have been another great uncle who became a missionary in Africa. The current Mr Berruyer also confirmed that his family had indeed held concerts in their apartment back in the day. Finally, on the table, they discovered Jean Romier's lighter.

The police investigation concluded that this was not a ruse pulled by Romier. The surprising details supplied by him about Mr Berruyer's family are, to say the least, a little disconcerting. Albert Einstein himself even seemed disturbed by the tale, and stated: "This young man fell through time... like tripping and falling down stairs."

The duration of the apparition – an entire evening – is exceptional in this genre, as is the fact that's it's not solely linked to the apartment on Rue de Vaugirard (because Romier claims to have first met Mr Berruyer in in the middle of the day in the Jardin du Luxembourg). If this were indeed true, then one can imagine it'd be possible to encounter such apparitions throughout the day and to be completely unaware of the fact – but we'll leave it at that.

The Premonitory Priest

Next, we find ourselves in Nantes, France in 1942. A woman in her forties rings the doorbell of Abbe Labrette, requesting that he come to hear the last confession of a dying young man. She supplies the address: 37, Rue Descartes, second floor. Labrette dutifully heads out, only to find the young man in perfect health. Both conclude that it was either a mistake or a prank. Labrette decides to hear the young man's confession anyway.

As he is leaving, Nantes is bombed and Labrette is forced to run to a bomb shelter. He volunteers to help at a medical centre later that evening. While giving absolution to the dying, he sees the corpse of the young man whom he'd visited several hours earlier. He looks in the young man's wallet and finds a picture of a woman closely resembling the one who rang at his door that evening. On the back he sees the inscription "Mama", written in the same handwriting as the address left to the priest. There are also two dates: 7 May, 1898 to 8 April, 1939.

Unlike the "Concert of Corpses" tale, in this instance there is a tangible piece of evidence associated with the events. There is also an aspect of intervention, which presumes the apparition possessed a certain form of knowledge pertaining to future events.

While apparitions of the deceased are hard enough to fathom, the existence of doppelgangers who replicate the living somehow seems even less explicable, in particular when their appearances are difficult to explain by way of trickery or hallucination. Our next story is a classic of this genre.

A Case of Bilocation

It's 1840 in Dijon, France. Like many of her peers, Émilie Sagée leaves Dijon to teach French in Russia. In 1845, she gets a comfortable position teaching in an institute for the daughters of nobility near Riga. Several weeks pass, and reports of panicked students begin to pile up.

Miss Sagée is frequently seen in two different locations. In one instance, a class of 20 students sees Sagée literally double before their eyes: one version of her is standing at the blackboard and explaining their text, while another, pale and transparent, is imitating her gestures beside her. The students, terrified, leave the building one by one.

Several days later, the school's 42 students are engaged in embroidery. When they look out of the window they can see one Miss Sagée picking roses in the garden, while another version of her remains seated silently in the classroom. One student attempts to touch the classroom version, and goes right through her.

The director receives threats to close the institution unless order is restored, and Sagée is dismissed from her post.

In this case, the number of witnesses and repeated sightings make fraud difficult. However, doppelgangers are not as uncommon as one might think. Various famous figures throughout history have even claimed to have had the privilege of meeting their doppelgangers, such as Guy de Maupassant (although here we must be cautious, as the timing roughly corresponds with the first symptoms of the paralysis that would claim his life) and Johann Wolfgang von Goethe, who also gives an account of the following apparition.

Goethe and Friedrich

Johann Wolfgang von Goethe takes a stroll in the countryside near Weimar one evening with a friend. Along the way, he's surprised to run into another acquaintance, Friedrich, who's oddly dressed in slippers and a bathrobe. When Goethe approaches to say hello, Friedrich suddenly disappears. Goethe becomes worried, wonders if he's gone crazy, and returns home to find Friedrich asleep beside the fire, wearing a robe and slippers. It transpires that Friedrich had stopped by unannounced and fallen asleep waiting for Goethe to return. During his brief nap, Friedrich also dreamed that he had walked down the road in his robe and slippers, intending to run into Goethe.

Here the credibility rather stems from Goethe's reputation, as well as those of the other people who witnessed the events.

We can, of course, go on to cite numerous other cases that are the subject of investigations, paranormal research and other forms of inquiry in Great Britain and the USA, but these mere four examples seem sufficient simply to call into question the overall plausibility of such events in a way that merits further consideration of their possible origins.

Dreams, Sleep, and Death: Schematic Interpretation

The case of Goethe seems to be a remarkable example of a unique sort of continuum between dreams and apparitions.

Dreams are extremely important according to our schema's interpretation of the various phenomena surrounding death. Sleep involves a loss of consciousness, which implies a relaxation of the connection between the superluminal consciousness and the cerebral cortex that's at the origin of accessing states of awareness or complete consciousness.

Numerous studies show that humans experience two forms of sleep: orthodox and paradoxical. Orthodox sleep represents the common notion of dreamless sleep during which EEGs show a state of complete calm. During paradoxical states, on the other hand, which last roughly a fifth of the night and are associated with dream states, we see a veritable explosion of activity: sleep is lighter, the heartbeat and respiration become irregular, blood pressure rises above normal, concentrations of adrenaline and cortisone in the blood rise rapidly, and brain temperature rises. People enter into paradoxical phases five or six times a night.

Studies also show that individuals who receive sufficient amounts of orthodox sleep, but are deprived of paradoxical sleep, quickly become ill and depressed, developing severe neuroses or psychoses that can prove fatal. However, sufficient paradoxical sleep, even when orthodox sleep is restricted, allows individuals to maintain their health. This state of awareness can be understood as a state of semi-consciousness: the superluminal conscious mind filtered through the cerebral cortex, such that only certain information passes through to be encoded and transformed into holograms.

In contrast, during orthodox sleep the cerebral cortex rests, thereby interrupting this function. The holograms disappear, and we see periods of dreamless sleep. However, as the total consciousness contains all the information and memory pertaining to the individual, it's possible for it to produce holograms periodically of a second type that are less consistent and more obscure than those which make up our daily reality.

This second type of hologram corresponds to paradoxical phases and dreaming, which originate in the total consciousness and can be assimilated into the category of visions that are seen in NDEs (such as wondrous landscapes and the appearances of loved ones), which serve to acclimatize the individual to the realm of the dead. In both cases, this acts as a sort of intermediate stage that precedes the superluminal domain, whose information and meaning are totally abstract.

This schema is also reminiscent of Greek mythology, which identifies Hypnos (sleep) as the twin brother of Thanatos (death), and the interpretation of modern physics seems to support the proximity of this relationship: when we dream, we stroll through death's waiting room.

Incidentally, this also explains premonitory dreams, as there is an implied access to the conscious domain's information pertaining to the individual's past, present and future. Future events can therefore manifest unintentionally in the form of a hologram of the second type while the individual is dreaming.

Fundamentally, then, our schema observes no physical difference between the appearance of the living or dead in either dreams or apparitions. Under the right conditions, holograms of the second type can be projected into our habitual spacetime, breaking out unexpectedly from the domain of the total conscious. The characteristic physical properties of apparitions (fleeting, transparent and immaterial) would imply that these are formed from particularly transient luminal particles.

Émilie Sagée's doppelgangers can then be understood to be the product of a kind of recurring fault in her holographic processing system, such as an illness affecting her superluminal consciousness and its interactions with the cerebral cortex. This resulted in a permanent hologram, called Émilie Sagée, superimposed upon another of the second type, which intermittently broke through the superluminal domain, thereby giving her students the impression of seeing two Émilies (with the second appearing faint and transparent).

In the case of Friedrich appearing to Goethe in the middle of the countryside in robe and slippers, there's a glaring continuity between Friedrich's dream and the apparition. Friedrich indeed dreamed that he was walking along the road and met Goethe at around the time when the latter was surprised by the doppelganger. This all fits loosely within our schema's description of holograms of the second type, which would consist of escaped elements native to the superluminal domain, and which are then projected onto normal subluminal spacetime (through Friedrich's dream state in this instance).

Apparitions of deceased persons occur by the same means. To comprehend this, we must recall that we've assumed – while necessarily relying heavily on NDE testimonies – that in the case of an irreversible death, the partial integration of the conscious into the superluminal domain is progressive and first must pass through a luminal phase. However, this phase can become permanent if the partial conscious hasn't amassed enough information to integrate itself into the total conscious of the superluminal domain.

Holograms of the second type are common in the luminal phase, some pleasant, some frightening, and are created by the deceased's state of mind. Therefore, it's possible for the deceased, in passing through, and all the more so if they remain there definitively, to project a hologram of the second type that produces a representation of themselves in our subluminal domain. Depending on the nature of this holographic projection, this would allow them to appear in dreams, where the deceased are represented very precisely, or apparitions that are observed one or more witnesses.

However, two cases of apparitions that we've discussed, "The Concert of Corpses" and "The Premonitory Priest", don't seem to fall into this category. They both involve apparitions that present certain physical characteristics and material properties. We can therefore deduce that these involved *true* subluminal holograms, indistinguishable from their subluminal counterparts. These holograms could only be produced by conscious entities that have crossed the light barrier and have access to superluminal spatiotemporal properties. Their integration into this system, which acts as a great river of superluminal world lines, may not yet be complete.

This would explain the interventionist character of the apparition appearing in the priest's premonition. The mother's consciousness could already benefit from access to superluminal spatiotemporal properties, including their complete world lines. To her, there was therefore no present, past or future, and she could perceive all the events of her son's life instantaneously. She could see the circumstances of her son's death and instantaneously (for her) project herself as a subluminal hologram into the hours leading up to his demise, thereby alerting the priest. The resulting materialistic hologram superimposed itself on what we call the present and, the deed accomplished, suddenly disappeared – which leads one to conjecture that there is another way to cross the light barrier besides death. The complexity of this method of analysis shows that it's important not to fall in the temporal trap. The mother of the young man had died three years prior, but for her this interval passed instantaneously, on account of her conscious mind already having transcended the subluminal domain.

In the case of "The Concert of Corpses", the egregious materialization of subluminal holograms apparently occurred without a precise goal. There was no obvious interventional goal, and the young man otherwise appeared to be in fine shape. This was a true and very material superposition of world lines: that of the young man, and those belonging to individuals from a past era.

Given that the superluminal consciousness has access to all the events of one's own subluminal life, it's reasonable to assume that it can access those of others as well. Normally, through the intermediary of partial consciousness, only the events comprising one's life are projected into the subluminal domain. These are then realized in a causal order dictated by the flow of time. However, just as in optical systems, the focus can be a little off. The subject therefore sees pieces of the past in their present – and even perhaps a bit of the future.

We'll make two additional remarks about "The Concert of Corpses".

First, the phenomenon was only witnessed by the young Romier. The neighbours witnessed nothing. This suggests the projection originated from the single superluminal conscious of this individual.

Second, if this concert really took place historically in the course of the lives of the deceased, this implies that Romier intervened in their past – but nothing proves that. Nothing would, as the superluminal conscious has the capacity to replace and substitute causal sequences instantaneously. One can then wonder whether or not there are multiple versions of the past, each corresponding to its own variation in causal sequences, and whether individuals voluntarily keep just the one in their memory, or if they're filtered out through the cerebral cortex.

In any event, when superluminal spacetime and consciousness come into play, the lesson is that Aristotelian logic needs to be abandoned for a new approach.

The Petit Trianon

On the subject of these phenomena, we can't go without mentioning the famous tale of the Petit Trianon.

Around the year 1900, two young English tourists were walking through a park in Versailles when, for several minutes, they were projected into the 18th century: they met and spoke with people from this era, in particular a young courtier who prevented them from continuing their stroll.

Our schema's interpretation is evidently the same. Although there's often talk of "hot spots" with regards to this sort of phenomena, it seems simpler to interpret this as the crossing of the world lines of these young men with certain members of the court of Marie-Antoinette. Einstein's comment about Romier – "This young man tripped and fell through time" – captures the heart of this interpretation.

Cases of apparitions therefore call for diverse explanations, but each can be interpreted through the schema of a superluminal domain and a new developing form of physics. Through objective consideration of these phenomena, we avoid losing important information to superstition and ridicule.

The diverse reports involving communication with the dead fall into this category.

Spiritualism

Spiritualism originated in 1847 in the hamlet of Hydesville, USA, and was enthusiastically promoted by the three Fox sisters. It's important to note that mid-19th-century

American Puritanism showed a particular fascination with magnetism and hypnotism: both could be readily witnessed for a relatively affordable fee and were thought to reveal the existence of hidden forces of human nature (an opinion shared by Victor Hugo, Maupassant and Balzac).

Spiritualism and automatic writing met with particular success between 1850 and 1920, flourishing amidst a mixture of positivism and charlatanism. It goes without saying that while some acts of spiritualism yielded intriguing results, much was revealed to be fraud fuelled by a public thirst for the sensational. At the same time, psychiatry, psychoanalysis and other methods were hard at work interpreting human behaviour as manifestations of the unconscious, repressed desires and even multiple personalities – which, on the whole, resulted in a body of approaches that had a notable penchant for inciting scepticism.

Jürgenson's Recordings

We'll end this chapter with one last example: Friedrich Jürgenson's 1959 tape recordings. Previously, Jürgenson had accidentally made an audio recording that revealed a voice resembling that of his mother who'd been dead for four years. He took to reproducing similar experiments, and wound up recording voices of the deceased in almost every language. Jürgenson continued this research with Friedbert Karger, a physicist at the Max Planck Institute in Munich, who verified the findings. The psychologist Konstantin Raudive was also able to reproduce similar results systematically.

The fact that these results were verified by numerous of physicists and engineers is compelling. But, coincidentally, for the phenomena to manifest, one or several people needed to be present.

The first explanation that comes to mind is that these are actual audio recordings from deceased individuals. Another would be to consider the information encoded within the memories of the individuals present at the recording, which would then serve to recreate the voices.

Here again, we see that a bit of caution is worthwhile; however, we can also humbly suggest that these recordings could be holograms projected from partial conscious minds stuck in the luminal state.

10

A NEW CONSCIOUSNESS AND REINCARNATION

The concept of reincarnation forms part of some of the most ancient religions in the world. Over 4,000 years before our present day, Hinduism embraced the idea that each person and animal is born, dies and reincarnates as a different life form according to the spiritual quality of the life they chose to live before (i.e. *karma*). This birth–death cycle, or *samsara*, continues indefinitely until the soul attains a degree of spiritual perfection known as *moksha*. In the 6th century BCE, Buddhism adopted a similar doctrine according to which the ultimate goal was to obtain *Nirvana*, a state of true self free from pain and desire where reincarnation would cease. There are other examples from other faiths as well.

Pythagoras and Plato

The first known Western allusion to reincarnation took the form of *metempsychosis* and comes to us from Pythagoras (although it's a good bet that he met some Easterners and borrowed the idea, which would have been new and original in his culture). Plato, who was in many ways his heir, took up the doctrine and added a punitive interpretation of reincarnation that reflected poor choices made throughout one's life, as illustrated by the Myth of Er in Book X of *The Republic* – one of literature's first NDEs.

Er is a soldier. He's left for dead on the battlefield. His soul leaves his body and arrives somewhere divine, where he witnesses other souls being judged. The evil are tortured, each receiving ten times the punishment for each fault, while the good enjoy bliss in paradise. He follows the souls destined for reincarnation toward a column of light where the Fates (Clotho, Lachesis, and Atropos) lay out their options:

> Each picks up one that falls near them. Everyone knows the ranks they can choose from. The hierophant spreads out many more choices than there are souls present, and there are many options: animals, humans, tyrants, some living until they die naturally, others whose lives are interrupted in the middle. [...] There were lives of famous men – whether for physique, beauty, strength, combat skills, noblesse or heritage. There were less extravagant choices. [...] But these possibilities didn't imply any determination regarding the character of the soul, as that would necessarily change following the decision at hand. [...]
>
> It's a sight to see, because it's pitiful, ridiculous and strange. Decisions are, in fact, often based on the manner in which someone led their former life.

Each is free to choose at will, but the choice is irreversible. Afterwards, the souls drink from the River Lethe, which causes them to forget their former life and experience in the afterlife. Er learns that his mission is to convey what he's seen to the living. He escapes and regains consciousness 12 days later to tell the tale.

Among the ancient Western religions, it's also interesting to note that one of the first Christian theologians, Origen of Alexandria (184–254 CE), believed in reincarnation. However, his ideas were quickly condemned by the Church, after which the idea of reincarnation remained effectively banished from Western thought.

Helena Petrovna Blavatsky: Theosophy

The 13th century, and in particular the 19th century, saw philosophers such as Kant, Hume and Schopenhauer propose serious hypotheses regarding reincarnation. The Western notion of reincarnation began to evolve again and found itself in the limelight once more through the works of Helena Petrovna Blavatsky (1831–1891), a follower of esoteric Buddhism and author of *Isis Unveiled* (1877) and *The Secret Doctrine* (1881). Blavatsky also founded the Theosophical Society, which increased awareness of these ideas in Western Europe.

Albert de Rochas: Hypnotic Regression

Between 1890 and 1913, Colonel Albert de Rochas, the director of the École polytechnique, began using hypnosis (which was all the rage in the 19th century) to assist people in recalling and visiting their past lives through hypnotic regression. This culminated in the 1911 publication of his book *Les vies successives* ("Successive Lives"), which contained 19 cases of what de Rochas believed to be actual instances of reincarnation. A word of caution, however, as there is a tendency for individuals to invent, more or less consciously, outlandish memories inspired by literature and fantasy.

Vladimir Raikov, Dennis Kelsey and Joan Grant

Nevertheless, hypnotic regression was later taken up by numerous psychiatrists in the 1950s and 1960s, notably Vladimir Raikov, Denis Kelsey and Joan Grant (Kelsy and Grant were married, and Grant was additionally a reputed medium and author of several books, including *Our Past Lives*). Their goal was to use hypnotic regression to encourage the emergence of a state of consciousness that was free of complexes, phobias and neuroses – some of which were attributed to experiences in past lives. For example, one individual had a particular complex pertaining to bird feathers, and this was associated with a trauma that occurred in a former life where they'd been left for dead in the desert and devoured by vultures.

Understanding the origin of their phobias allowed individuals to liberate themselves from them, which has an obvious therapeutic value, regardless of one's viewpoint. Psychodrama, which is often employed in group therapy, seems somewhat comparable.

Ian Stevenson: the Scientific Study of Reincarnation

The year 1966 saw the first attempt to examine reincarnation scientifically, with the publication of the book *Twenty Cases Suggestive of Reincarnation* written by American psychiatrist Ian Stevenson after he had conducted 15 years of research. Stevenson looked at over 100 cases worldwide, which naturally were reported more frequently in regions that believed in the phenomena, such as India, Sri Lanka, Lebanon, Brazil and Alaska.

Interestingly enough, most cases involved children who reported experiencing memories of past lives. Statistically, this phenomenon generally began between the ages of two and three, reaching a peak at ages eight to ten, and then progressively diminishing (sometimes disappearing entirely at around 20 years of age). Stevenson points out that – with the exception of a single case in Lebanon – the manifestation of these memories exclusively involved individuals between the ages of 5 to 15.

This may be suggestive of memory deformation (a well-studied phenomenon), but more importantly is problematic given the nature of taking testimonies from children. Stevenson also encountered a significant language barrier, as he spoke neither the requisite Arabic nor Hindi to record certain testimonies directly.

Stevenson was therefore very understandably cautious about publishing his conclusions, which otherwise naturally needed to defend their credibility against a variety of other criticisms as well. In face of accusations of fraudulent testimony, Stevenson emphasized that any publicity the parents received would be detrimental, in particular when there were socioeconomic differences with an alleged former family that could lead to custody battles. He also drew on socioeconomic differentials and geographic locations to refute claims of cryptomnesia (in the sense that testimonies could involve a subconscious attribution of currently accessible information to the memory of a past life). Stevenson was also forced to defend his research against more extravagant claims, such as ESP and possession, which he did with more simple common sense, arguing that if the individual giving the testimony could read people's minds, why wouldn't they exercise their faculties more fully rather than merely limit themselves to collecting memories of a single deceased individual. With possession, he simply pointed out that there was no overlap between the two lifespans in question.

Outside of rebuttal, Stevenson began characterizing certain patterns observed in his research. First, some children had talents that couldn't be explained as the sudden manifestation of hereditary gifts. For example, one young man (who claimed to be a reincarnated woman) manifested a talent for sewing at age three. Another three-year-old, who claimed to be a former lemonade salesman, built a complex machine to make the drink. Stevenson also found that birth marks were indicative and cites scars on the neck of a child who claimed to recall a past life in which they'd either been strangled or their throat had been slit. Finally, he notes certain trends toward recognizing locations that held a unique significance to the alleged former incarnation.

The Case of Swarnlata

The first case we'll look at is Swarnlata, who was born on 2 March, 1948, in Shapur. Stevenson became aware of the case in 1961 and researched it in detail.

Swarnlata was the daughter of an employee at the School Inspector's Office in the district of Chhatarpur. She belonged to a moderately comfortable socioeconomic class and Stevenson noted that her family was very cultured (for example, the entire family also spoke English, and Stevenson could interview Swarnlata directly without need for interpreters).

Ever since she had been three years old, Swanlata's family began noticing that she claimed to be recalling memories from a past life where she was named Biya Pathak. On one occasion, while passing through the village of Katni with her father, roughly 200 miles from their home, she claimed to be very close to where she lived in her former life.

As the years went by, Swanlata found herself recalling very complex chants and dances that weren't accessible to her through her family or school. In 1958, at the age of nine, she recognized a former neighbour in Katni and addressed them by name.

In 1959, her parents became intrigued and began to look more into Katni. They took a trip there with their daughter, who immediately began pointing out her past home and recognizing members of her former family, including her brother and nephews. She described compelling details about the layout of the house, its former condition, the social status of the very wealthy family and events of her former life – in which she married young, had two sons and died in

1939 from a heart attack. The members of the Pathak family were quickly convinced and didn't hesitate to acknowledge her as their departed loved one.

Stevenson returned to visit Swarnlata in both 1971 and 1973 to see how the situation evolved, which was exemplary: Swanlata obtained a master's degree in botany and was teaching in Chhatarpur. She married in 1973 and kept up relations with her former family, who admired her.

Swarnlata's case is both classic and frequently mentioned, especially in countries that believe in reincarnation and don't ostracize individuals who claim to experience such memories. However, this remains a vast and underexplored area of research, where further analysis and quantification would be of great benefit.

The Case of Rose-Mary Adrian and Michel Davel

The case of Michel Davel is both much wilder and was more thoroughly investigated. It can be thought of as a temporary reincarnation of a deceased personality within a living adult.

On 11 November, 1918, a penniless young French sailor called Michel Davel attended a ball in Calais to celebrate the recent armistice. There he met Rose-Mary Adrian, a particularly beautiful, intelligent and wealthy young woman. The two fell madly in love, but Mr Adrian thought Davel was much too poor to marry his daughter and refused to give his consent.

In 1919, the Adrian family left France and settled in Australia. Rose-Mary's parents later passed away, leaving her alone in Melbourne. One day in 1934, she ran into Michel on a street in the city. He explained that he'd been in a car accident on 12 August of that year, was experiencing a bit of amnesia, and that he'd recently arrived in Australia. To Rose-

Mary, Michel had changed a bit physically, but his memories of Calais remained intact and he still carried his love for her. Several months later they got married, and Michel was appointed to an important position in a business owned by family friends of the Adrians.

Thirteen years passed. Then, in 1947, Michel disappeared for several days without warning. When he returned, he spoke impeccable English where before he had a thick French accent. He said that his memories had returned from before his car accident in 1934, that his name was actually George Littlon, and that he'd lived in Adelaide prior to his accident and was married. His former wife, whom he contacted, instantly recognized him as George Littlon, a devout Australian who'd never left the country.

So Rose-Mary Adrian left Australia to live in England. She investigated what had happened to the real Michel Davel and found that he'd died in a serious accident on 12 August, 1934 – the same day when, thousands of miles away, George Littlon had been in a coma following a serious accident.

Multiple sources verified both that Michel Davel had never visited Australia and George Littlon had never left it, which ruled out the chance that the two had ever physically met. No elements of fraud were otherwise identified on the part of either party.

The only explanation to turn to was, then, extraordinary: the reincarnation of Davel's soul, at the moment of his death on 12 August, 1934, into the body of George Littlon, which had been rendered unconscious on account of an accident. If this is indeed possession, one of the first questions that comes to mind is what became of the real Littlon during this time, from 1934 to 1947. That's difficult to answer. Why would Davel's conscious eclipse it for 13 years and then suddenly cede its place? Such are the insurmountable obstacles encountered from the vantage point of our current understanding.

Reincarnation Vis-à-vis Our New Physical Schema of the Conscious Mind

It's possible in the future that more sophisticated technology and research will allow a more thorough investigation into reincarnation. Whatever the case may be, for the moment we can at least attempt an interpretation, in part through our new physical schema of the conscious mind, and in part from a distinct temporal logic that stems from relativistic theory.

In 1971, the Gerhard R Steinhauser published a remarkable work entitled *The Chrononautes*, which was one of the first attempts to explain life after death using modern physics (he even makes an allusion to the light barrier). Several pages are dedicated to analysing reincarnation, where Steinhauser makes several interesting points:

> The world's population has exploded over the past decade, leaving a current population of over three billion individuals. This then raises certain questions. Where are all these souls coming from? Is there an endless supply? Should we believe that none have ever been formerly incarnated on Earth?[1]

He goes on to cite parapsychic research (in particular involving mediums) that offers many explanations for the existence of memories of past lives, such as those described in Stevenson's research, and notably states:

> And to conclude, if we assume that life continues after death, why would these "souls" necessarily incarnate on Earth rather than some other star or planet? This is more suggestive of the fact that the theory of reincarnation dates back to an epoch when the Earth was still considered to be the centre of the universe and there was no conception of the possibility of life throughout the cosmos.[2]

Steinhauser argues that we need to revise reincarnation theories in light of modern cosmological advances, in particular our awareness of the now billions of galaxies, each of which plays host to billions of stars: even the most conservative estimates suggest that there should be millions of planets with forms of intelligent life that are vastly different from life on Earth. There's then a need to reframe reincarnation in a larger cosmological context that encapsulates the vastness of the known universe.

Another difficulty reincarnationists encounter is duration. Classically, souls are considered to pass from life to life and from century to century (that is, according to the flow of subluminal time) in a long chain. Doesn't it seem a tad tedious to engage in an incessant wandering that stretches for hundreds of thousands of years and is destined to involve such an inordinate amount of pain and suffering?

Steinhauser is again one of the first to offer a different interpretation:

In 1960, the newspapers printed a story about a brave Bavarian ironworker who claimed to have lived, during a series of dreams that took place in a single night, a life as a knight in a castle during the Middle Ages. The man was able to confirm details regarding his past life with historical records.[3]

This story, which is not unique in its nature, might until now fall under the category of resurrection or reincarnation. Our current schema, however, puts us on a different trail. The Bavarian could act as a sort of temporal variable, whose conscious mind simultaneously directed two lives on different temporal planes – which is also an alternative explanation for similar cases involving reincarnation. So do multiple lives exist? Do we carry with us pieces of former and future lives that we become conscious of at one time or another?

Steinhauser's idea is absolutely remarkable and concurs with the conclusions that we've drawn from our own new schema for the conscious mind. Given that the superluminal spacetime is not subject to the flow of time, the present, past and future become simultaneous. Any event, whatever the date, is instantaneously accessible to the superluminal consciousness. It's only through projection into the subluminal that events are filtered and organized into causal sequences and linked to the flow of time, thereby giving the impression of a past, present and future.

Nothing we've postulated stands in the way of the superluminal conscious being capable of projecting information pertaining to multiple subluminal lives. One life could occur in ancient Athens, for example, while another is in medieval France: perhaps a third in 18th-century USA, a fourth in 1980s Germany, and a fifth in the 25th century in some other country. A single superluminal conscious mind would be linked to all these lives, each of which correspond to a collection of information realized in a given moment.

In our normal state, and through the filtering of the cerebral cortex that allows us to create and focus on what we call the present, each individual feels that they live a determinate life that plays out during a specific epoch. But this is a virtual reality, and they could in fact be living many different lives at different epochs, which would be *simultaneous* from the superluminal perspective.

This would then explain certain phenomena pertaining to pseudo-past lives. For some individuals, who generally tend to be young according to Stevenson's research, the filtration engine running between the superluminal conscious and the cerebral cortex hasn't been completely tuned up, resulting in an interference of simultaneous lives (one or many) and the impression that one of these is *past* or *former*.

This would explain many of Stevenson's cases, as well as many others involving memories of the same genre. As the superluminal conscious mind remains constant, we would see similarities between its various subluminal projections, and hence similar traits and characteristics between the lives involves (such as Stevenson's birthmarks, gender, etc.).

The case of Rose-Mary Adrian and Michel Davel is less textbook: a simultaneity of lives from the superluminal perspective that translates into an extraordinary subluminal continuity. At the moment of Davel's death, his conscious and cerebral cortex cease to interact. At the same time, George Littlon, who was in a coma following brain trauma, experiences a similar cessation. Then somehow their wires get crossed and Davel's consciousness finds itself leaving the luminal domain, en route to the momentarily vacant cortex of George Littlon. Littlon, who likely experienced an NDE during his coma, then spends 13 years in the luminal light barrier (a duration he would have perceived as infinitesimal according to luminal spacetime). Their wires then get uncrossed, and Littlon suddenly return to his normal self 13 years after the event.

If one appeals to the spirit of romance, there's also a more audacious interpretation: Davel's recently deceased conscious mind, upon arriving in the luminal domain, recognizes and instantaneously views the panoramic of his and Rose-Mary Adrian's destiny, and chooses to take advantage of a momentary lapse to commandeer Littlon's body. In any case, according to Stevenson's criteria, Davel's situation was not a case of reincarnation, but rather a form of possession.

Steinhauser's brave Bavarian, on the contrary, is a more classic example of reincarnation. However, instead of manifesting in a young child, a fault in the cerebral cortex resulted in its manifestation in an adult. With age, and a more developed defence against superluminal information, the aspects of his simultaneous lives (the past, from a terrestrial point of view) came to him instead through dreams, which we've already seen are closely linked to the superluminal domain. Dreams are in fact a method of communication, albeit imperfect, between the subluminal and superluminal domains, because they occur when the normal interaction between the conscious mind and cerebral cortex is at rest.

Death, visions, dreams and reincarnation are nothing more than diverse examples of phenomena that manifest from the superluminal domain and project into our subluminal world. These are doors that we don't quite know fully how to use to allow us to cross through the light barrier.

The total conscious mind is like white light, and the light barrier a prism, where passage through it results in the simultaneous decomposition and distribution of seven colours into multiple subluminal lives that appear across a succession of time. We thereby glimpse the inner workings of both reincarnation and multiple personalities, and the various hints and signs revealed through hypnosis and certain conditions of hysteria.

We carry within us, without knowing it, the history of humanity. We are the witnesses of millions of years past. We are living archives. Each one of us represents our own set of global information that comprises the destiny of the universe.

CONCLUSION

By way of conclusion, we'd like to return to some of the main ideas that have defined our discussion. First, we'd like to emphasize that the schema we've proposed is merely a philosophical attempt founded on the extrapolation of a purely physical theory. One wouldn't attribute any definitive scientific value to its propositions, which can certainly be modified and improved, in particular as our understanding progresses.

We began with some ideas that seemed to arise naturally from Régis Dutheil's research in physics. On the whole, physics has started to demonstrate a growing acceptance of the idea that the universe isn't limited by our observable surroundings and that there's perhaps a part of reality that escapes our senses and understanding. With matter comprising one of our fundamental criteria for what's real, the discovery of new particles should alter our conception of reality itself and bring about new schemas: new ways of perceiving and understanding our surroundings.

That's what we've tried to pursue in the present work, in the wake of the theoretical postulation of the existence of tachyons (which we hope will be followed by experimental verification).

These particles constitute a new genre of matter, different to that which we know, and with surprising new properties. Time no longer flows. Observers can potentially have access to instantaneous knowledge on all the events of their life.

This other matter, this other reality, nevertheless remains imperceptible to our eyes. This mystery, combined with the nature and reality of the conscious mind, drives us to share our schema which, we repeat, is nothing more than a hypothesis intended to inspire further research.

Our attempt to endow the conscious mind with a material reality is, in fact, nothing new. We've witnessed it many times over, in a variety of schemas set forth by prominent names in science in recent decades. What's new is the association of the conscious mind with a different form of matter and the use of physics to resolve questions that lie at the intersection of philosophy and science.

Our goal, in this work, is precisely to present a multidisciplinary approach that unites physics, medicine, philosophy and history, against questions that are often only seen in fragments.

The implications of superluminal properties of conscious matter (the equivalence between space and time, the absence of a flow of time, instantaneity, decreasing entropy and disorder – or, equivalently, the constant augmentation of information and meaning) are as much philosophical as they are physical. Plato's dualist vision of the universe is not too far off from where we now find ourselves, which is also very close to the ideas of David Boehm and Karl Pribram. This superluminal or conscious domain would be fundamental, a point of origin from which our subliminal world is projected like a hologram. There would be a constant correspondence between the superluminal spacetime of total information and its holographic representation in the subliminal domain. For example, the destiny of a human being, from birth to death, would arrange itself in the superluminal domain according to affinities between elements of information, which would be projected, according to causal and temporal sequences, onto our subliminal perception. The cerebral cortex would filter this incoming flux, blocking our perception of the whole and focusing on the information pertinent to our surrounding environment.

If we reflect further, this schema also constitutes, as we've seen, a solution to the difficult problem of destiny versus free will, which has preoccupied philosophers and theologians for centuries.

From states of consciousness to the reality of the universe, to the mystery of death, we've taken a small step toward understanding the conscious mind before and after the physical body dies.

Since we are nothing more than holographic projections of our superluminal conscious, we can take advantage of memories that we have from when we're unconscious, such as dreams and near-death experiences, to probe further into the nature of the superluminal domain and the concept of a time that doesn't flow. In turning to the plethora of testimonies systematically collected over the past 15 years by American academics, we've shown that our schema offers unique interpretations of such phenomena and, in some cases, actual testimonies of near-death experiences.

The future lies with synthesizing our approaches, and the systematic analysis and testing of new schemas and ideas. And while this future is always seen as separate on our subluminal plane, each step we take brings us closer and contributes to the edification of the great fountain of human knowledge.

NOTES

1. Conceptualizing Consciousness

1 See Spinoza's introduction to *Ethics*: "By mode I understand the affections of a substance, or that which is in another through which it is also conceived."

2 Contemporaries of Leibniz.

3 Sometimes called the Realm of Forms.

4 See the allegory of the cave in Plato's *Republic*, Book VII.

5 Philosophers who preceded Socrates.

6 Here we see that the soul has gained a certain imminence on the Platonic ideal, as it's accorded no existence outside the matter whose development it provokes. Regardless, Aristotle emulates Plato by dividing the soul into three parts: the *static*, the *sensory* (the principles of organic functions and instinct in plants and animals), and the *rational* (unique to humans). We therefore have a basis of quasi-biological observations layered with metaphysical reflections, and a heterogeneity that is arguably more complex than Plato's –even more so as Aristotle also incorporates a third and fourth cause.

7 See *Meditations on First Philosophy* and *Passions of the Soul*.

2. Consciousness and Reality

1 James Clerk Maxwell incorporated optics through the identification of light as electromagnetic waves and condensed the majority of known phenomena at the time into extrapolations of four fundamental equations (known as Maxwell's equations).

2 e:energy in joules, v: the frequency of the radiation, h: Planck's constant.

3   *A reference frame* describes the relative motion and coordinate system of an observer. The speed of light being constant in all reference frames means that no matter the motion of an observer with respect to a beam of light, light is always to be measured as travelling at the same speed of 300,000 km/s – which is decidedly different from what happens if we measure the speed of a car, a plane, or virtually any other object.

4   The 1881 Michelson and Morley experiment is famous for both its brilliant original design and anomalous results. The intention was to reveal minute variations in the speed of light that would arise on account of the Earth's motion through the *ether*: the hypothetical medium in which 19th-century physicists postulated electromagnetic waves (hence light) propagated. Their measurement process takes advantage of the Earth's orbit around the Sun (which averages 30 km/s in the direction of orbit) and the wave nature of light. A single beam of coherent neon light was split and set along two equidistant branches: one perpendicular to the Earth's motion, another parallel to it. The Earth's motion through the ether would then result in a differential in the duration of the beam through these passages (a spatial offset as well, which can be shown to be negligible), which would be observable as variations in the *interference pattern* produced when the two beams recombined. However, after all the trails, reorientations, repetitions at different times of the year, and so on and so forth, Michelson and Morley never observed *any* variation in their interference patterns. They concluded that some other effect was at play, and it took the genius of Einstein to interpret these results much more simply: a proof by contradiction that the premise of their original experiment was incorrect, for there is no *ether*, and the speed of light is always to be measured as a constant 300,000 km/s regardless of the reference frame from which it is observed.

5   The emission of electrons induced by electromagnetic radiation such as light hitting a material.

6   The principle of complementarity.

7   These are encoded into the Ψ function mentioned above for mathematical representation.

8 An observable property that represents a degree of freedom associated with a photon's polarization.

9 In Europe, the European Centre for Nuclear Research (or CERN: *Centre européen de recherches nucléaires*) figures prominently.

10 *An Interpretation of Nature and the Psyche.*

11 Jung was also strongly influenced by Plato's theory of ideas.

12 Michel Cazenave, *La synchronicité, l'âme et la science,* Payot, 1984.

13 Emmanuel Swedenborg (1688–1772), the famous Swedish mathematician and philosopher. From 1743, he claimed to have visions and declared himself to be in contact with the spiritual world. Theosophist.

14 Recorded in *Dreams of a Spirit-seer* by Emmanuel Kant.

3. Living in a Material World

1 Eccles, *Facing Reality*, 1970, pp.118–19. Available at: https://www.informationphilosopher.com/solutions/scientists/eccles/

2 Ibid.

3 Ibid.

4 Quoted in "The Psychotropic Universe", Stargate, CIA reading room website. Available at: www.cia.gov/readingroom/document/cia-rdp96-00792r000400330013-4

5 That is, faster than the speed of light.

6 A more tangible example of an interference pattern is what you get after dropping several stones onto the surface of a calm body of water: a series of ripples with peaks and troughs propagating in circular patterns that interfere constructively (when two peaks or troughs combine to make a bigger version) or destructively (when they cancel each other out). Freezing the pattern on the surface would be analogous to recording diffused interference patterns on holographic film, from which one would be able to reconstruct the images of the stones.

7 According to the Fourier theorem, any mathematical function can be considered as the sum of an infinite number of sinusoidal functions whose frequencies vary infinitesimally. Fourier analysis consists of applying this process, which decomposes a given function into its basic pieces, either mathematically or experimentally.

4. Proposing a New Schema for the Conscious Mind

1 That is, faster than the speed of light.

2 The branch of mathematics known as *group theory* shows that there exists a *group* corresponding to the subluminal Lorentz transformations, and a second group, isomorphic to the first, corresponding to the superluminal Lorentz transformations.

3 The actual 3D *spatial* + 1D *time* analogy of this behaves similarly, although it is more difficult to visualize on account of the extra dimension.

4 See K. Pribham, *La synchronicity et le fonctionnement du cerveau* ("Synchronicity and Brain Function") in *La Synchronicity, l'Ame et la Science* by H. Reeves et al, (Payot, 1984), p. 111.

5 The reader will find a mathematical exposé on the superluminal schema of consciousness, considered as a field of tachyonic matter in the article of R Dutheil and B Dutheil: "A new temporal model: synchronicity and acausality in the superluminal universe," *Revue internationale de Biomathématique* (1987), 1st trimester, n.97.

  Elsewhere the exhaustive exposé of the theoretical works of R Dutheil on tachyonic physics can be found in the work *Théorie de la Relatvité et Mécanique quantique cans la région du genre espace* (Editions Derouaux, Liège, Belgium), which also contains references on the communications of R. Dutheil.

6 The experimental approach to tachyons has entered a new phase: since 1980, Professor Jacques Steyaert of the Nuclear Physics Institute at the University of Louvain-La-Neuve has pursued this goal. Using the institute's large cyclotron, he observed a new mode of interaction between gamma rays and matter, electrons in particular. These experiments involved the production of a pair of particles that he interpreted as tachyons. These tachyons would be magnetic monopoles, capable of producing electric current. J. Steyaert gave the name *tachyoelectric effect* to the phenomena he discovered, and research is underway to confirm this observation. R. Dutheil and J. Steyaert presented a report on magnetic monopoles with respect to the tachyoelectric effect to the 1989 session of the International Congress on General Relativity in Barcelona: "The Dirac equation in two rectilinear dimensions leading to magnetic monopoles in the

light cone coordinates" (*Scientific World*). We note that the work of the French theoretician Prof. Georges Lochak on magnetic monopoles is absolutely fundamental in this domain.

5. A Brief History of Death

1   For a more precise account, see Stuart Edelstein's *Biologie d'un mythe* (Sand, 1988).
2   Translated from Philippe Ariès, *Essais sur l'histoire de la mort en Occident du Moyen Age à nos jours,* le Seuil, Paris, 1975.
3   Ibid., p.30
4   Ibid., p.37
5   Statistics confirmed by a survey taken by the IFOP and published in *Le Monde* on 10 January 1986.

6. Near-Death Experiences: the Research

1   In particular: *The Dying Patient as a Teacher: An Experiment and An Experience* (1965), *Questions and Answers on Death and Dying* (1974), and *Death, the Final Stage of Growth* (1975).
2   Raymond Moody, *Life After Life*, p.30.
3   M Sabon, pp. 26–27.
4   M Sabon, p.25.

7. Near-Death Experiences: Individual Testimonies and Our Schema

1   Raymond Moody, *Life After Life*, p.42.
2   Kenneth Ring, *Memories of Death*, p.92.
3   Moody, pp. 45–46.
4   Ring, p.98.
5   Ibid.
6   Moody, p.51.
7   Moody, p.55.
8   Ring, p.56.
9   Moody, pp.54–55.
10 Moody, p.63.
11 Moody, p.60.
12 Moody, p.68.

13 Ibid., pp.68–69
14 Ibid., p.69.
15 Ring, p.101.
16 Ibid. pp. 101–102.
17 Ibid.p.102.
18 Ibid.
19 Moody, pp.75–76.
20 Moody, p.82.
21 Moody, p.83.
22 Ring, p.60.
23 Ring, p.63.
24 Ring, p.64.
25 Pribram made the contention that the conscious mind, without the brain knowing, is capable of projecting holograms from interference patterns.
26 Moody, pp.85-87.
27 Moody, p.88.
28 Moody, p.89.
29 Moody, p.90.
30 Ring, p.77.
31 The interference patterns would in fact be wave interference in superluminal phases or packets of subluminal waves translated by the brain to the conscious in the form of holograms.
32 Moody, p.101.
33 Ibid.
34 Ring, p.105.
35 Sabom, p.85.

8. Our Schema and the Need for a New Physics to Understand Death

1 Arthur C Clarke, *2001: A Space Odyssey*, p.168.

10. A New Consciousness and Reincarnation

1 Gerhard R Steinhauser, *The Chrononautes*, p.46.
2 Ibid.
3 Ibid., p.106.

WATKINS
1893

The story of Watkins began in 1893, when scholar of esotericism John Watkins founded our bookshop, inspired by the lament of his friend and teacher Madame Blavatsky that there was nowhere in London to buy books on mysticism, occultism or metaphysics. That moment marked the birth of Watkins, soon to become the publisher of many of the leading lights of spiritual literature, including Carl Jung, Rudolf Steiner, Alice Bailey and Chögyam Trungpa.

Today, the passion at Watkins Publishing for vigorous questioning is still resolute. Our stimulating and groundbreaking list ranges from ancient traditions and complementary medicine to the latest ideas about personal development, holistic wellbeing and consciousness exploration. We remain at the cutting edge, committed to publishing books that change lives.

DISCOVER MORE AT:

www.watkinspublishing.com

Read our blog

Watch and listen to
our authors in action

Sign up to
our mailing list

We celebrate conscious, passionate, wise and happy living.
Be part of that community by visiting

/watkinspublishing @watkinswisdom
/watkinsbooks @watkinswisdom